The Logica Yearbook
2019

The Logica Yearbook 2019

Edited by

Igor Sedlár
and
Martin Blicha

© Individual authors and College Publications 2020
All rights reserved.

ISBN 978-1-84890-327-2

College Publications
Scientific Director: Dov Gabbay
Managing Director: Jane Spurr

www.collegepublications.co.uk

Original cover design by Laraine Welch

All rights reserved. No part of this publication may be reproduced, stored in a retrieval system or transmitted in any form, or by any means, electronic, mechanical, photocopying, recording or otherwise without prior permission, in writing, from the publisher.

Preface

This volume contains a selection of the contributions presented at the conference *Logica 2019*, which took place at the Hejnice Monastery in northern Czech Republic on June 24–28, 2019.

The long tradition of Logica symposia started in 1987. Logica welcomes both mathematical and philosophical logicians, and one of its main aims is to foster a fruitful exchange of ideas between these overlapping groups. The informal atmosphere aims to encourage a dialogue between logicians of all generations, including students. As the editors of this volume, we are proud to contribute to the successful completion of the annual symposium cycle by presenting this collection to you.

Logica 2019 was, just as all the previous Logica symposia, organized by the Department of Logic of the Institute of Philosophy of the Czech Academy of Sciences. More than thirty lectures were presented during the conference, including those given by our distinguished invited speakers David Makinson, Per Martin-Löf, Alessandra Palmigiano and David Ripley. A tutorial was given by Ansten Klev. Unfortunately, these post-proceedings can offer only a limited record of the topics discussed and cannot hope to even partially convey the symposium atmosphere. Nevertheless, we believe that the articles in this volume can more than stand on their own.

Both the Logica symposium and The Logica Yearbook are the result of a joint effort of many people to whom we would like to express our gratitude. We are, of course, very grateful to the Institute of Philosophy of the Czech Academy of Sciences for all the support that made the event possible. We would also like to thank the staff of Hejnice Monastery for their hospitality and friendly assistance. We gratefully acknowledge the support of the Bernard Family Brewery of Humpolec, the traditional sponsor of the Logica social programme. We owe thanks also to the Czech Science Foundation, which provided significant support for the meeting and for the publication of this book with funding from the grant project no. 17-15645S, and to the Czech Academy of Sciences for support within the Strategy AV21 programme. Ondrej Majer designed the trademark Logica T-shirts. Olga Bažantová proved again to be a key member of the organizing team by dealing with administrative and practical matters efficiently and with wit. We are of course grateful also to College Publications and its managing director, Jane Spurr, for our consistently pleasant cooperation during the preparation of each Logica Yearbook Series volume. Last but not least, we would like to thank all of the authors for their contributions and collaboration during the editorial process.

Prague, May 2020

Igor Sedlár and Martin Blicha

Table of Contents

A Degree-Theoretic Framework for Feasible Knowledge 1
  Libor Běhounek

Dialogical Criteria of Adequate Formalisation 17
  Simon Brausch

Relative Interpretations and Substitutional Definitions of Logical
Truth and Consequence 33
  Mirko Engler

The Problem of Harmony in Classical Logic 49
  Giulio Guerrieri and Alberto Naibo

Expressing Validity: Towards a Self-Sufficient Inferentialism 67
  Ulf Hlobil

Logic and Ethics 83
  Per Martin-Löf

Revision Operator Semantics: Some Order for the Zoo of
Conditionals 93
  Eric Raidl

Strong Normalization in Core Type Theory 111
  David Ripley

Negation on the Neo-Australian Plan 131
  Sebastian Sequoiah-Grayson

Alethic Pluralism and Logical Consequence 147
  Nicholas J.J. Smith

Proof-Theoretic Semantics and the Interpretation of Atomic
Sentences 163
  Preston Stovall

Cut-free and Analytic Sequent Calculus of Intuitionistic Epistemic
Logic 179
  Youan Su and Katsuhiko Sano

Do We Need Recursion? 193
  Vítězslav Švejdar

A Classical Bimodal Logic with Varying Essences 205
    Andrew Tedder
Categoricity and Possibility. A Note on Williamson's Modal
Monism 221
    Iulian D. Toader

A Degree-Theoretic Framework for Feasible Knowledge

LIBOR BĚHOUNEK[1]

Abstract: A semantics for epistemic logic is defined that retains the epistemic agent's logical rationality while avoiding logical omniscience. The proposed solution to the logical omniscience paradox is based on resource-awareness implemented by means of t-norm fuzzy logics.

Keywords: epistemic logic, logical omniscience, feasibility, t-norm fuzzy logic, possible-world semantics

1 The logical omniscience paradox

Standard epistemic logic (see, e.g., Meyer, 2003) renders the epistemic modality *the agent knows that* φ as a propositional modal operator K. The operator is supposed to validate some of the standard axioms of normal modal logics, which formalize several principles of epistemic reasoning:

(K)	$\vdash K(\varphi \to \psi) \to (K\varphi \to K\psi)$	logical rationality
(T)	$\vdash K\varphi \to \varphi$	truth of knowledge
(4)	$\vdash K\varphi \to KK\varphi$	positive introspection
(5)	$\vdash \neg K\varphi \to K\neg K\varphi$	negative introspection
(Nec)	$\varphi \vdash K\varphi$	necessitation

The axioms (4) and (5) of positive and negative introspection are often considered optional, depending on the introspective abilities of the epistemic agent. The doxastic variant with the modality *the agent believes that* φ replaces the truth axiom (T) by the consistency axiom (D):

(D)	$\vdash K\neg\varphi \to \neg K\varphi$	consistency of beliefs

---
[1] The work was supported by project No. LQ1602 "IT4I XS" in the program NPU II of the Ministry of Eductation, Youth and Sports of the Czech Republic.

Libor Běhounek

Being a normal modal logic, standard epistemic (or doxastic) logic is sound and complete with respect to the usual Kripke semantics, where K is evaluated □-like and the intended meaning of the accessibility relation R is that of epistemic indistinguishability: in a possible world w, the agent only knows (or believes) that the actual world is one of the worlds w' such that wRw', but cannot distinguish which one of these.

A well known defect of standard epistemic logic is the paradox of *logical omniscience*, or the fact that by the inference rule (Nec) and the axiom schema (K), the agent automatically knows all propositional consequences of their actual knowledge, including all propositional tautologies. This is, of course, unrealistic for real-world agents.

Since the standard axioms and rules of doxastic logic likewise include (K) and (Nec), the problem of logical omniscience applies to standard doxastic logic as well. We will therefore deal with both of these variants at once, for the most part referring to epistemic logic only, but *mutatis mutandis* applying our considerations to doxastic logic too.

2 Solutions based on resource-awareness

The logical omniscience paradox is obviously caused by over-idealization of deductive abilities of epistemic agents in standard epistemic logic. In particular, the modal axioms neglect the costs of logical derivations, or the fact that the agent needs to spend some of their limited resources, such as computation time, memory, or energy, in order to perform the derivation. Consequently, an important class of solutions to the logical omniscience paradox is based on *resource-awareness*, i.e., on modifying the standard account by acknowledging that some resources need be spent on logical deductions.

Several resource-aware solutions to logical omniscience have been proposed, i.a., by Duc (1997, 2001) and Artemov and Kuznets (2014). One of these proposals is to introduce time-awareness by prepending the temporal modality $\langle F \rangle$ (standing for "at some future point") to the conclusions of epistemic principles. The axioms (K) and (4) are thus modified to read:

$$\vdash K(\varphi \to \psi) \to (K\varphi \to \langle F \rangle K\psi)$$
$$\vdash K\varphi \to \langle F \rangle KK\varphi,$$

meaning: "if the agent knows both φ and $\varphi \to \psi$, then *at some future point* (namely, after applying modus ponens), the agent will know ψ" and "if φ is

A Degree-Theoretic Framework for Feasible Knowledge

known, then *at some future point* (namely, after performing introspection), $K\varphi$ will be known"; and similarly for the axiom (5) and the rule (Nec).

The simple temporal modality $\langle F \rangle$ can further be refined into dynamic modalities whose programs represent deduction steps: e.g., composing the programs $\langle mp \rangle$ for modus ponens and $\langle pi \rangle$ for positive introspection, the appropriately modified axioms (K) and (4) yield:

$$\vdash K(\varphi \to \psi) \to \big(K\varphi \to \langle mp; pi \rangle KK\psi\big)$$

Another approach to resource-awareness consists in syntactic stratification of the epistemic modality K, e.g., by counting the steps needed for the logical derivation. The stratified axioms (K) and (4) then read:

$$\vdash K^n(\varphi \to \psi) \to (K^m\varphi \to K^{n+m+1}\psi)$$
$$\vdash K^n\varphi \to K^{n+1}K^n\varphi,$$

and similarly for the stratified axiom (5) and rule (Nec).

All of the aforementioned solutions avoid logical omniscience by deferring the knowledge deduced from the agent's actual knowledge to some future point. An unsatisfactory feature of these solutions, though, is that they solve the paradox not by treating the modality K itself, but by substituting some modification thereof, such as $\langle F \rangle K$ or K^n. The aim of the present paper is to solve the logical omniscience paradox by a resource-aware treatment of the very modality K. We will achieve this goal by distinguishing the actual, potential, and feasible knowledge of an epistemic agent and employing suitable non-classical logics for resource-aware reasoning about the costs of logical derivation.

3 The logics of costs and resources

As has been argued by the present author in one of the previous volumes of *The Logica Yearbook* (Běhounek, 2009), most kinds of typical resources exhibit the structure of a *(semi)linear residuated lattice*.[2] In more detail, it can be observed that many common kinds of resources come to us in amounts that can be compared and added or subtracted. The comparability establishes a linear lattice order on amounts; the addition and subtraction of amounts then constitute the residuated lattice's operations of fusion and residual. Admittedly, in some of the more complex types of resources (imagine, for

---

[2]In algebraic parlance, *semilinear* means subdirectly decomposable into linear components.

instance, the lists of cooking ingredients in recipe books), the order may no longer be linear. Nevertheless, even such resources can often be decomposed into some components (oil, flour, sugar, salt, etc.) whose amounts are fully comparable, and so the order is at least semilinear.

The logics that are sound and complete with respect to the algebraic semantics of (semi)linear residuated lattices are known as *fuzzy logics*.[3] Thus, arguably, logics that are generally suitable for resource-aware reasoning, epistemic or otherwise, can be found among fuzzy logics.[4]

Before we proceed, let us recall a few properties of propositional fuzzy logics (for more details see, e.g., Hájek, 1998; Běhounek, Cintula, & Hájek, 2011). In this paper, we will only need the following facts:

1. The standard set of truth values for fuzzy logics is the real unit interval $[0, 1]$. Propositional connectives are interpreted truth-functionally, by certain well-behaved operations on $[0, 1]$. Particular fuzzy logics differ in the choice of these truth functions for connectives.

2. The prominent fuzzy logics G, Ł, and Π interpret conjunction (or *fusion,* denoted by $\otimes$) by the following truth functions on $[0, 1]$:

    Gödel–Dummett logic G: $\quad \|\varphi \otimes \psi\| = \min(\|\varphi\|, \|\psi\|)$
    Łukasiewicz logic Ł: $\quad \|\varphi \otimes \psi\| = \max(0, \|\varphi\| + \|\psi\| - 1)$
    Product fuzzy logic Π: $\quad \|\varphi \otimes \psi\| = \|\varphi\| \cdot \|\psi\|$

3. These three logics, as well as all of their relatives from the family of so-called *t-norm fuzzy logics,* validate this law for implication:

    $$\|\varphi \to \psi\| = 1 \quad \text{iff} \quad \|\varphi\| \leq \|\psi\|$$

Let us now give a brief description of the cost-based interpretation of t-norm fuzzy logics; for more details see the original paper (Běhounek,

---

[3] The delimitation of the class of fuzzy logics as the logics of classes of (semi)linear residuated lattices is due to Cintula (2006).

[4] By virtue of having the algebraic semantics of residuated lattices, fuzzy logics fall within the broader family of *substructural logics* (see, e.g., Ono, 2003; Kowalski & Ono, 2010). A substructural logic that is often regarded as the logic of resources is *linear logic.* However, as argued in the previous paper (Běhounek, 2009), using linear logic for resources disregards the semilinear structure manifested by most kinds of resources—in other words, neglects their decomposability into components with linearly ordered amounts. Consequently, linear logic as the logic of resources is unnecessarily weak, and under-generates resource-wise, compared to suitable fuzzy logics.

2009). Under this interpretation, the truth values of fuzzy logic represent *costs,* or amounts of a particular resource, such as money, time, or memory. The designated truth value 1 represents zero costs and all other truth values represent non-zero costs; the smaller the truth value, the larger the cost. The truth functions of fuzzy logic perform certain operations on costs. For instance, the operation of fusion $a \otimes b$ yields the value of the two costs a and b put together, while the implication $a \to b$ expresses the surcharge to a that would yield at least the cost b; the remaining propositional connectives of fuzzy logic possess a natural meaning in terms of costs as well. Particular fuzzy logics differ in the manner of combining costs: e.g., the way costs are combined by the standard truth function for $\otimes$ in Łukasiewicz logic Ł is that of *bounded additivity* (where 0 represents the maximal cost); in product fuzzy logic Π, *unbounded additivity* (via the logarithm, with 0 representing an infinite cost); and in Gödel–Dummett logic G, *maxitivity* (as is appropriate, e.g., for various kinds of capacity). The formulae of fuzzy logic then represent various ways of combining costs assigned to atoms.[5] Specifically, tautological implications $\varphi \to \psi$ express valid laws of cost preservation, of the form: "The cost ψ never exceeds the cost φ." Fuzzy logics can thus be viewed as calculi that formalize reasoning *salvis expensis,* in a similar manner as classical logic formalizes reasoning *salva veritate.*[6]

In the next sections I propose a solution to logical omniscience based on resource-aware reasoning modeled in fuzzy logic. The approach has already been briefly sketched in an earlier short paper (Běhounek, 2013). The present paper elaborates the solution in more detail, taking in part advantage of the recently developed fuzzy intensional semantics (Běhounek & Majer, 2018). Due to limited space, most of the mathematical apparatus is omitted here and deferred to a later full exposition.

---

[5] Formulae thus directly represent combinations of costs, rather than propositions or states of affairs. Compare this *formulae-as-costs* interpretation, e.g., with *formulae-as-types* in the Curry–Howard correspondence or the categorial grammar interpretation of the Lambek calculus. Alternatively, the logical atoms can be interpreted as gradual propositions of the form "the item x is inexpensive"; however, that requires an additional assumption on the correspondence between combining costs and combining degrees of truth by the propositional connectives.

[6] The general algebraic semantics of t-norm fuzzy logics in terms of semilinear residuated lattices generalizes this interpretation even to resources with non-linearly ordered amounts (yet decomposable into linearly ordered components).

4 Three kinds of knowledge

When discussing solutions to logical omniscience based on resource awareness, it is generally useful to distinguish the following three kinds of a given agent's knowledge:

- *Actual knowledge,* or the explicit knowledge that is immediately available to the agent. In artificial agents, this can be the contents of their database. In humans, it is the sum of all the facts the person knows without needing to infer them from other known facts.

- *Potential knowledge,* or the implicit knowledge that is, at least in principle, logically derivable from the actual knowledge. In other words, the set of all logical consequences of the agent's actual knowledge.

- And *feasible knowledge,* or that part of potential knowledge that the agent can feasibly derive from actual knowledge, taking into account the agent's limited resources (such as time, memory, etc.).

It can be noticed that logical omniscience is only troublesome for the notion of *feasible* knowledge, since the *actual* knowledge of a non-idealized real-world agent is never closed under logical consequence (one reason being its finiteness); while any agent's *potential* knowledge does indeed include all logical truths by definition.

It can furthermore be observed that whereas actual knowledge can be viewed as crisp and finite (and potential knowledge as crisp and infinite), feasible knowledge is apparently *gradual,* since long and complex logical derivations will require more of the agent's limited resources (e.g., time, memory, or energy) than shorter or simpler ones—and so can be *less feasibly* performed by the agent. The fact that fuzzy logics are, by design, tailored to deal with gradual notions just further underscores their suitability for resource-sensitive treatment of feasible knowledge.

As a matter of fact, the paradox of logical omniscience can be construed as an instance of the sorites paradox: a *single* additional step of logical derivation is always feasible and within the capability of a logically rational agent.[7] However, just like in the sorites paradox, it does not follow that arbitrarily long logical derivations would be feasible, with however many steps. In consequence of this correspondence, any solution to the sorites paradox generates a solution to the logical omniscience paradox. Our proposed solution

---

[7]Except when hard limits (e.g., on time or memory) are imposed. This case can be set aside, though, as then neither the sorites series of steps nor logical omniscience arise.

to logical omniscience thus, besides making use of resource-awareness, also parallels the treatment of the sorites paradox in fuzzy logic (cf. Hájek & Novák, 2003).

Combining these ideas, we will render feasible knowledge as a fuzzy set of formulae that are logically derivable from the actual knowledge, where the membership degrees represent the derivation costs. The next sections elaborate the fuzzy semantics of feasible knowledge in more detail.

5 Fuzzy modal logic of feasible knowledge: syntax

Following the previous considerations, we are going to formalize feasible knowledge as a unary fuzzy modality K governed by a suitable t-norm fuzzy logic $\mathcal{L}$. As explained in Section 3, the choice of the fuzzy logic depends on the way we intend to compound costs (for example, Łukasiewicz logic is suitable for bounded addition).

The epistemic modality K can, in principle, be applied to formulae of any language $\mathcal{E}$ in which the agent's knowledge is formalized.[8] Our syntax will, therefore, be two-layered, allowing:

(i) The modality K to be applied to a formula φ in the language $\mathcal{E}$ of the agent's knowledge representation, and

(ii) Modal atoms of the form Kφ to be combined by the connectives of a propositional t-norm fuzzy logic $\mathcal{L}$ of our choice.

We will denote the logic with this two-layered syntax by $\mathcal{L}_K[\mathcal{E}]$. For example, if the agent's knowledge is represented in first-order multi-agent doxastic logic KD45$\forall_B^n$ with unary modalities $B_1, \ldots, B_n$ and unary predicates P, Q, then $K\bigl((\forall x)B_1(Px \to B_n Qx)\bigr) \otimes K\bigl((\exists y)\neg B_1 Py\bigr)$ is a well-formed formula of the logic Ł$_K$[KD45$\forall_B^n$].

It can be noticed that in $\mathcal{L}_K[\mathcal{E}]$, the modality K of feasible knowledge cannot be nested. This is because the 'outer' logic $\mathcal{L}$ serves for *our* reasoning about the agent's feasible knowledge, and its formulae are not part of the agent's knowledge. It is, nevertheless, possible that the language $\mathcal{E}$ of the agent's reasoning contains its own epistemic modality that refers to the agent's own knowledge: for instance, the agent might use the standard epistemic logic S5$_K$ for their epistemic reasoning. Strictly speaking, the

---

[8]The language is usually equipped with a set of derivation rules the agent can apply, i.e., a logic of the agent's reasoning. We will therefore regard $\mathcal{E}$ rather as a logic than just its language.

agent's ('inner') epistemic modality and our ('outer') epistemic modality K should always be denoted by different symbols. Nevertheless, since they are distinguished by the level of nesting ('our' K's being the outermost ones), in logics such as Ł$_K$[S5$_K$] we will take the liberty of using the same symbol K on both syntactic levels. This will make our syntax closer to that of standard epistemic logic and allow us, for example, to write positive introspection in its traditional form K$\varphi \to$ KKφ.

Besides nesting, the two-layered syntax of $\mathcal{L}_K[\mathcal{E}]$ also prohibits combining modal and non-modal atoms by the connectives of $\mathcal{L}$. This restriction appears natural, as the connectives of $\mathcal{L}$ are intended to just combine the costs of knowledge. It is, nevertheless, reasonable to relax this constraint and permit forming mixed formulae too. One reason is that non-modal atoms can as well be employed to denote costs (as, e.g., in Section 8 below), and then they become meaningfully combinable with the costs of knowledge. Moreover, the costs themselves can be regarded as the degrees of truth of certain propositions (e.g., when we interpret Kφ as the graded statement *"to infer φ from the actual knowledge is inexpensive"*), and thus as freely combinable with any other graded propositions of the logic $\mathcal{L}$. So, provided that $\mathcal{E}$-formulae can be assigned truth values of $\mathcal{L}$ (as is the case, e.g., if the logic $\mathcal{E}$ is bivalent, since all $\mathcal{L}$-algebras contain the truth values 0 and 1), we will also consider the extension of $\mathcal{L}_K[\mathcal{E}]$ that admits mixed formulae and denote it by $\mathcal{L}_K(\mathcal{E})$. This will allow us, for instance, to discuss the truth axiom K$\varphi \to \varphi$ within the framework of such logics as Ł$_K$(S5).[9]

6 Fuzzy modal logic of feasible knowledge: semantics

We will introduce our cost-sensitive fuzzy modal logic $\mathcal{L}_K[\mathcal{E}]$ of feasible knowledge by means of a fuzzy-relational possible-world semantics, where possible worlds represent the agent's epistemic states. Each possible world is assigned a set of $\mathcal{E}$-formulae that form the agent's actual knowledge in that state. Transitions between the states correspond to inference steps the agent is able to perform; i.e., to changes of the actual knowledge in consequence of deductions performed by the agent. As usual, possible transitions are encoded by an accessibility relation; in our case, the relation is weighted ($\mathcal{L}$-valued, or fuzzy) and the weights represent the costs of the transitions.

A sample model for $\mathcal{L}_K[\mathcal{E}]$ is depicted in Figure 1. The agent's possible

---

[9] If $\mathcal{E}$ contains its own epistemic modality, then $\mathcal{L}_K(\mathcal{E})$ needs, unlike $\mathcal{L}_K[\mathcal{E}]$, to distinguish both modalities graphically, in order to disambiguate such formulae as KK$\varphi \to$ Kφ.

A Degree-Theoretic Framework for Feasible Knowledge

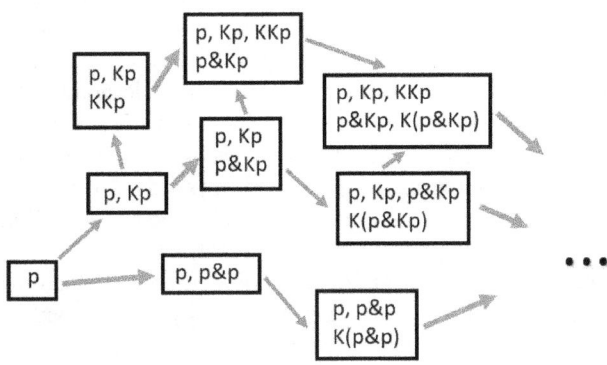

Figure 1: Example of an epistemic model.

epistemic states (represented by boxes) differ in the agent's actual knowledge (written inside the boxes). For simplicity, let us assume that the logic $\mathcal{E}$ of the agent's reasoning has only the rules of adjunction and positive introspection. Possible transitions between the states, brought about by application of these deduction rules, are indicated by arrows. The width of each arrow signifies the cost incurred by the agent for performing the inference step: in this model, applying adjunction is less costly (hence, thicker arrows). Collectively, the arrows represent the cost-weighted accessibility relation between the epistemic states.[10]

Let us have a look at how the agent's feasible knowledge is computed for any given epistemic state in Figure 1. Suppose, for instance, that the actual world w_0 is the leftmost one, in which the agent's actual knowledge consists just of the single $\mathcal{E}$-formula p, and let us calculate the degree to which the formula $\mathrm{K}(p \,\&\, \mathrm{K}p)$ is part of the agent's feasible knowledge in that state; in other words, we want to determine the truth value of the $\mathcal{L}_\mathrm{K}[\mathcal{E}]$-formula $\mathrm{K}\big(\mathrm{K}(p \,\&\, \mathrm{K}p)\big)$ in the state w_0.[11] It can be observed that in this model, the

---

[10]The figure omits the arrows that arise by composition of the depicted single-step arrows; their costs can be calculated as the $\mathcal{L}$-conjunction (fusion) of the costs of the single-step transitions. Also omitted are the full-width (i.e., zero-cost) arrows from each state to itself. These all, too, are part of the graded accessibility relation, which equals the $\mathcal{L}$-valued reflexive-transitive hull of the weighted single-step arrows from the picture (cf. Section 7).

[11]Recall that the outermost K is the $\mathcal{L}_\mathrm{K}[\mathcal{E}]$-modality governed by the fuzzy logic of costs $\mathcal{L}$,

$\mathcal{E}$-formula $\mathrm{K}(p \mathbin{\&} \mathrm{K}p)$ belongs to the agent's actual knowledge just in the two upper-rightmost depicted states. We can see that in order to arrive at this knowledge (i.e., at one of these states), the agent needs to perform positive introspection twice and adjunction once. The cost of deriving this knowledge is thus the fusion of the costs of these three inference steps. Consequently, the truth value of $\mathrm{K}\bigl(\mathrm{K}(p \mathbin{\&} \mathrm{K}p)\bigr)$ in the state w_0 is the $\mathcal{L}$-conjunction $\alpha \otimes \beta \otimes \alpha$, where α is the degree representing the cost of positive introspection (or the weight of the thinner arrows in Figure 1) and β the cost of adjunction (the thicker arrows) in our $\mathcal{L}_\mathrm{K}[\mathcal{E}]$-model of Figure 1.

It can be noticed that in this $\mathcal{L}_\mathrm{K}[\mathcal{E}]$-model, there are also other (longer) paths from w_0 to a state containing $\mathrm{K}(p \mathbin{\&} \mathrm{K}p)$. These paths consist of three applications of positive introspection and one adjunction, and so are no cheaper than the shorter route described above. Since we are interested in resources that must *unavoidably* be spent to arrive at the knowledge, we take the cheapest route, or generally the infimum of costs (which is the supremum of degrees) along all possible paths.[12]

Models for $\mathcal{L}_\mathrm{K}(\mathcal{E})$, which admits mixed formulae, will only need to additionally set an $\mathcal{L}$-evaluation of $\mathcal{E}$-formulae in each state. Let us summarize these ideas in a definition of fuzzy epistemic models.

Definition 1 *A fuzzy epistemic model for $\mathcal{L}_\mathrm{K}(\mathcal{E})$ is a tuple* $\mathbf{M} = (W, \mathbf{L}, R, A, e)$, *where:*

- W *is a non-empty set (of an agent's possible epistemic states);*

- $\mathbf{L}$ *is an $\mathcal{L}$-algebra (of truth degrees representing costs);*

- $R \colon W^2 \to \mathbf{L}$ *is an $\mathbf{L}$-valued weighted accessibility relation on W (representing the transition costs between states);*

- $A \colon W \times \mathit{Form}_\mathcal{E} \to \{0,1\}$ *is a relation indicating the agent's actual knowledge* $A^w = \{\varphi \in \mathit{Form}_\mathcal{E} \colon A(w,\varphi) = 1\}$, *in each state w; and*

---

while the inner modalities K belong to the logic $\mathcal{E}$ of the agent's reasoning.

[12] Admittedly, this is a gross idealization, as *finding* the cheapest route would generally be a non-trivial problem for the agent. (Thanks are due to Sebastian Sequoiah-Grayson for pointing this out at *Logica 2019*.) We adopt the idealization here, since—unlike the resource-obliviousness of standard epistemic logic—it does not produce logical omniscience. Indeed, a more realistic model of cost-aware epistemic reasoning ought to address it—possibly by replacing the costs of moving along the cheapest path with the costs of *searching* for whichever path to the desired knowledge. Then, besides the inferential distance in the logic $\mathcal{E}$, the feasibility degree would also depend on the agent's path-searching heuristics and algorithms.

A Degree-Theoretic Framework for Feasible Knowledge

- $e\colon W \times Form_{\mathcal{E}} \to \boldsymbol{L}$ is an evaluation of $\mathcal{E}$-formulae in each state.[13]

The truth value $\|\psi\|_w$ of an $\mathcal{L}_{\mathrm{K}}(\mathcal{E})$-formula ψ in a state $w \in W$ is defined by the following Tarski conditions, for all n-ary $\mathcal{L}$-connectives c, $\mathcal{E}$-formulae φ, and $\mathcal{L}_{\mathrm{K}}(\mathcal{E})$-formulae $\psi_1, \ldots, \psi_n$:

$$\|\varphi\|_w = e(w, \varphi)$$
$$\|c(\psi_1, \ldots, \psi_n)\|_w = c^{\boldsymbol{L}}(\|\psi_1\|_w, \ldots, \|\psi_n\|_w)$$
$$\|\mathrm{K}\varphi\|_w = \bigvee\nolimits_{w' \in W,\, \varphi \in A^{w'}} Rww'$$

The notions of intension $\|\psi\|\colon w \mapsto \|\psi\|_w$, tautologicity and (local) entailment with respect to a class of fuzzy epistemic models are defined in a standard manner.[14]

7 Constraints on fuzzy epistemic models

To achieve some level of real-world plausibility, fuzzy epistemic models need be constrained by some conditions on the weighted accessibility relation that would reflect various principles of the agent's epistemic reasoning. Some of such constraints are the following:[15]

1. Fuzzy transitivity of R:

$$\forall w, w', w'' \in W\colon Rww' \otimes Rw'w'' \leq Rww''.$$

This condition reflects the concatenability of $\mathcal{E}$-derivations: if an $\mathcal{E}$-derivation δ_1 takes the agent from an epistemic state w to a state w' at the cost Rww' and a derivation δ_2 makes the transition from w' to w'' at the cost $Rw'w''$, then the transition from w to w'' costs at most $Rww' \otimes Rw'w''$, or the cost of the concatenated derivation $\delta_1\delta_2$.

---

[13]Note that although both e and A assign values to $\mathcal{E}$-formulae, their roles are different: the value $e(w, \varphi)$ determines the (degree of) *truth* of φ in w, while $A(w, \varphi)$ indicates whether φ is part of the agent's actual *knowledge*. We assume that e, though taking values in $\boldsymbol{L}$, is also an admissible evaluation of formulae in the sense of $\mathcal{E}$ (cf. the end of Section 5). In fuzzy epistemic models for the logic $\mathcal{L}_{\mathrm{K}}[\mathcal{E}]$, which does not admit mixed formulae, this component of a model is simply omitted.

[14]As usual in t-norm fuzzy logics, only the truth degree 1 designated in $\boldsymbol{L}$. Notice that in infinite models, the existence in $\boldsymbol{L}$ of all the suprema required by the Tarski condition for K need be assumed. This can be ensured either by requiring the lattice-completeness of $\boldsymbol{L}$, or (to enable axiomatizability) by using *safe* models (cf. Hájek, 1998, ch. 5).

[15]The properties of fuzzy transitivity, fuzzy reflexivity, upper sets, preimages, and closures are well studied in fuzzy set theory (e.g., Bělohlávek, 2002; Běhounek, Bodenhofer, & Cintula, 2008). Fuzzy relations that are fuzzy transitive and fuzzy reflexive are called fuzzy preorders.

2. Fuzzy reflexivity of R (i.e., $\forall w \in W \colon Rww = 1$), which expresses the immediate (cost-free) availability of actual knowledge.

3. Persistence, or the upperness of $A_\varphi = \{w \in W \colon A(w, \varphi) = 1\}$ in R for each $\mathcal{E}$-formula φ.

In fuzzy epistemic models that satsify these constraints, feasible knowledge can be characterized as a closure operator on the epistemic state space:

Observation 1 *Let* $\mathbf{M} = (W, \boldsymbol{L}, R, A, e)$ *be a fuzzy epistemic model for* $\mathcal{L}_\mathrm{K}(\mathcal{E})$ *satisfying the conditions 1–3 above and let φ be an $\mathcal{E}$-formula. Then:*

- *The intension $A_\varphi \subseteq W$ of the actual knowledge of φ is a crisp upper set in the $\boldsymbol{L}$-valued fuzzy preorder R on W.*

- *The fuzzy intension $\|\mathrm{K}\varphi\| \colon w \mapsto \|\mathrm{K}\varphi\|_w$ of the feasible knowledge of φ is the fuzzy preimage of A_φ in R. In symbols, $\|\mathrm{K}\varphi\| = R \leftarrow A_\varphi$.*

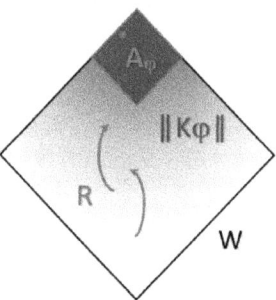

Figure 2: The fuzzy intension $\|\mathrm{K}\varphi\|$ of the feasible knowledge of φ is the fuzzy preimage of the (crisp upper) intension A_φ of the actual knowledge of φ in the fuzzy accessibility preorder R; i.e., $\|\mathrm{K}\varphi\| = R \leftarrow A_\varphi$.

Moreover, since R is a fuzzy preorder, the fuzzy preimage operator $X \mapsto R \leftarrow X$, where $(R \leftarrow X)(w) = \bigvee_{w' \in W}(Rww' \otimes Xw')$ for all $X \colon W \to \boldsymbol{L}$ and $w \in W$, satisfies the conditions of a fuzzy closure operator.[16]

---

[16]Namely, is pointwise extensive and monotone w.r.t. $\leq^{\boldsymbol{L}}$, and $R \leftarrow (R \leftarrow X) \doteq R \leftarrow X$ for all $X \colon W \to \boldsymbol{L}$. Additionally, $R \leftarrow \emptyset = \emptyset$. The (easy) proofs can be found, e.g., in (Běhounek et al., 2008).

A Degree-Theoretic Framework for Feasible Knowledge

Consequently, the feasible knowledge $\|\mathrm{K}\varphi\|$ can also be described as the (backward) fuzzy closure of the actual knowledge A_φ under the cost-weighted fuzzy transition relation R on epistemic states.

The list of constraints 1–3 above is neither unalterable nor complete. Reasonable conditions on fuzzy epistemic models depend largely on the assumed properties of the epistemic agent, and particularly on the logic $\mathcal{E}$ of the agent's reasoning. For example, persistence of actual knowledge (condition 3) can hardly be required if $\mathcal{E}$ is non-monotonic. On the other hand, possible additions may include the following constraints:

- The finiteness of A^w for all $w \in W$ (finite actual knowledge in each epistemic state).

- Pointwise inclusion[17] of the actual knowledge A in the model's $\mathcal{E}$-evaluation e (facticity of actual knowledge); or in doxastic variants, pointwise inclusion of A in any $\mathcal{E}$-evaluation $e'\colon W \times Form_\mathcal{E} \to L$ (consistency of actual beliefs).

- Constant evaluation of $\mathcal{E}$-formulae across all epistemic states, i.e., $e(w, \varphi) = e(w', \varphi)$ for all $w \in W$ and $\varphi \in Form_\mathcal{E}$ (static facts).[18]

- The property of (weighted) *confluence* for transitions between epistemic states, which reflects free combinability of derivations (if valid for $\mathcal{E}$):

  $\forall w, w_1, w_2 \in W \; \exists w_3 \in W\colon$
  $A^{w_3} = A^{w_1} \cup A^{w_2}, Rww_2 \leq Rw_1w_3, Rww_1 \leq Rw_2w_3.$

- If each $\mathcal{E}$-inference step produces a single $\mathcal{E}$-formula, the models are bound to satisfy the following condition for each $w \in W$ and $n \in \mathbb{N}$:

  $$\bigvee\nolimits_{w' \in [w]_{n+1}} Rww' \leq \bigvee\nolimits_{w' \in [w]_n} Rww',$$

  where $[w]_n = \{w \in W\colon \mathrm{Card}(A^{w'} \smallsetminus A^w) = n\}$.

---

[17] I.e., $A(w, \varphi) \leq e(w, \varphi)$ for all $w \in W$ and $\varphi \in Form_\mathcal{E}$.

[18] Notice that if $\mathcal{E}$-evaluation is state-dependent as in Definition 1, the truth-axiom $\mathrm{K}\varphi \to \varphi$ need not be valid in a model even if A is pointwise included in e, since the evaluation of φ can change during derivations.

8 The costs of inference steps

The general requirements on fuzzy epistemic models listed in Section 7 abstract from specific inferential abilities of epistemic agents. Still, it may sometimes be desirable to have the agent's capability of performing particular inferential steps (such as modus ponens or positive introspection) captured axiomatically, in a similar way as in standard epistemic logic. Lest we fall back into the logical omniscience paradox, though, the modal axioms of standard epistemic logic need to be adapted to reflect the agent's cost of inference. For instance, the cost-sensitive modifications of the epistemic axioms (K) and (4) might read:

(K′) $\quad \vdash K(\varphi \to \psi) \otimes K\varphi \otimes m_{\varphi,\psi} \to K\psi$
(4′) $\quad \vdash K\varphi \otimes i_\varphi \to KK\varphi$

Here, $m_{\varphi,\psi}$ and i_φ are new propositional constants added to $\mathcal{L}$, possibly different for each $\varphi, \psi \in \textit{Form}_\mathcal{E}$ (as the costs of inference steps generally depend on the formulae involved—e.g., on their length).

The appropriate values of these propositional constants in a model are contingent on the properties of the particular agent, and for many agents (such as humans) can hardly be determined precisely. Nevertheless, independently of their exact values, their presence in the axioms ensures that the costs of inference steps are accounted for. In longer $\mathcal{E}$-derivations, these costs accumulate by the fusion connective $\otimes$ of $\mathcal{L}$, eventually making too long derivations from the actual knowledge infeasible. This eliminates logical omniscience while keeping the agent inferentially capable, by virtue of the cost-sensitivity of the axioms of logical rationality in $\mathcal{L}$.

References

Artemov, S., & Kuznets, R. (2014). Logical omniscience as infeasibility. *Annals of Pure and Applied Logic, 165*, 6–25.

Běhounek, L. (2009). Fuzzy logics interpreted as logics of resources. In M. Peliš (Ed.), *The Logica Yearbook 2008* (pp. 9–21). College Publications.

Běhounek, L. (2013). Feasibility as a gradual notion. In A. Voronkov et al. (Eds.), *LPAR-17-short: Short Papers for 17th LPAR* (Vol. 13 of EasyChair Proceedings in Computing, pp. 15–19).

A Degree-Theoretic Framework for Feasible Knowledge

Běhounek, L., Bodenhofer, U., & Cintula, P. (2008). Relations in Fuzzy Class Theory: Initial steps. *Fuzzy Sets and Systems, 159*, 1729–1772.

Běhounek, L., Cintula, P., & Hájek, P. (2011). Introduction to mathematical fuzzy logic. In P. Cintula, P. Hájek, & C. Noguera (Eds.), *Handbook of Mathematical Fuzzy Logic* (Vol. I, pp. 1–101). College Publications.

Běhounek, L., & Majer, O. (2018). Fuzzy intensional semantics. *Journal of Applied Non-Classical Logics, 28*, 348–388.

Bělohlávek, R. (2002). *Fuzzy Relational Systems: Foundations and Principles* (Vol. 20). Kluwer/Plenum.

Cintula, P. (2006). Weakly implicative (fuzzy) logics I: Basic properties. *Archive for Mathematical Logic, 45*(6), 673–704.

Duc, H. N. (1997). Reasoning about rational, but not logically omniscient, agents. *Journal of Logic and Computation, 7*, 633–648.

Duc, H. N. (2001). *Resource-Bounded Reasoning about Knowledge* (Unpublished doctoral dissertation). University of Leipzig.

Hájek, P. (1998). *Metamathematics of Fuzzy Logic*. Kluwer.

Hájek, P., & Novák, V. (2003). The sorites paradox and fuzzy logic. *International Journal of General Systems, 32*, 373–383.

Kowalski, T., & Ono, H. (2010). Fuzzy logics from substructural perspective. *Fuzzy Sets and Systems, 161*, 301–310.

Meyer, J.-J. C. (2003). Modal epistemic and doxastic logic. In D. Gabbay & F. Guenthner (Eds.), *Handbook of Philosophical Logic* (2nd ed., Vol. 10, pp. 1–38). Kluwer.

Ono, H. (2003). Substructural logics and residuated lattices—an introduction. In V. F. Hendricks & J. Malinowski (Eds.), *50 Years of Studia Logica* (Vol. 21, pp. 193–228). Kluwer.

Libor Běhounek
University of Ostrava, CE IT4Innovations–IRAFM
The Czech Republic
E-mail: `libor.behounek@osu.cz`

Dialogical Criteria of Adequate Formalisation

SIMON BRAUSCH[1]

Abstract: By introducing dialogical logic to the theory of formalisation, this paper aims at restoring the inferential and holistic features of Peregrin and Svoboda's approach to adequately formalising natural language. The present investigation on the yet unexplored potential of what the dialogical approach can contribute to the question of formalisation will then amount to the development of genuinely inferential and holistic criteria of adequacy which offer a new perspective on the problem of formalisation.

Keywords: logical formalisation, adequacy criteria, dialogical logic, inferentialism

1 Introduction

Peregrin and Svoboda (2017, pp. 69–74) have recently proposed *inferential* adequacy criteria which take a *holistic* perspective on the problem of formalisation. However, the resulting theory fails to meet the underlying requirements of inferentialism and holism and thereby undermines the project's initial ambitions. The aim of this paper is thus to restore the inferential and holistic features of their approach by applying current developments in the field of dialogical logic to the theory of formalisation. Since the dialogical framework shares some of the most distinctive assumptions of Peregrin and Svoboda's understanding of logic and language, I claim that the resulting *dialogical* theory of adequate formalisation can avoid the problems mentioned above while staying consistent with their overall approach.

After a brief survey of Peregrin and Svoboda's take on the adequacy of formalisations, I will first propose a dialogical reformulation of their two inferential criteria and address the problem concerning holism. I will then

---

[1] I would like to thank Timm Lampert, who supervised the seminar paper this short essay is based on, as well as Zoé McConaughey for discussing various technical questions on dialogical logic. I am also grateful to Shahid Rahman, Jaroslav Peregrin, Vladimír Svoboda, and Friedrich Reinmuth for comments on this topic.

substitute their non-inferential auxiliary criteria with a dialogical *Principle of Correspondence* between formal and informal argumentative strategies. In the last section, I will illustrate how my dialogical approach can be put into practice by applying the *Principle of Correspondence* to an example.

2 Peregrin & Svoboda's Take on Adequate Formalisation

Peregrin and Svoboda (2017) develop their inferential theory of adequate formalisation in their most recent work *Reflective Equilibrium and the Principles of Logical Analysis*. As they are working within a *pragmatic-linguistic* conception of logic, they conceive logic to be *rooted in* and therefore *depending on* the inferential structure of natural language argumentative practice (2017, pp. 11–12). Concerning the meaning of language and the role of logic, they thus advocate *semantic inferentialism* (2017, p. 122) as well as *logical expressivism* (2017, p. 26). As a consequence, the authors wish not to take into consideration any semantic or syntactic features for assessing the adequacy of a logical formalisation. Instead, they develop two *inferential* adequacy criteria, the *Principle of Reliability* (REL) and the *Principle of Ambitiousness* (AMB), which focus on a sentence's *inferential behaviour*—and that of its formalisation—in a certain number of sample arguments in which the sentence (S) and the formalisation (Φ) occur (2017, pp. 70–71). While the first principle (REL) requires that a *valid* argument form will not yield natural language instances that are intuitively *in*correct, the second principle (AMB) keeps track of the opposite direction: if the sample argument is intuitively *correct* but its formalisation with Φ as premise or conclusion is *not* valid, then the argument form fails the test of *Ambitiousness*. In short, Peregrin and Svoboda's criteria require a *bilateral* correspondence between intuitively correct arguments and valid argument forms.

Now the crucial point is that the two principles taken together merely tell us if a given formalisation counts as adequate or not. However, this does not rule out the possibility of ending up with *several* adequate formalisations of the same sentence. And in such a case, one cannot identify the *more* adequate formalisation by means of their two inferential criteria. For this reason, the authors further introduce two *non*-inferential auxiliary criteria, the *Principle of Transparency* and the *Principle of Parsimony* (2017, pp. 72–73). For making a conclusive decision, one must then choose between two, sometimes opposing, syntactic considerations. Peregrin and Svoboda's approach for assessing the adequacy of formalisations is thus not *entirely* inferential. While their two primary criteria (REL) and (AMB) do represent this fundamental feature, the two auxiliary criteria are not at all in line with the requirements of an inferentialist theory of formalisation. The incapacity of their two inferential criteria to identify the *more* adequate alternative among several *adequate* formalisations leads us to the second problem, namely their insensitivity to syntactic features. I claim that a genuinely inferential theory of adequate formalisation should not have to rely on non-inferential criteria but still be capable of taking into account the syntactic peculiarities which do in fact play an important role with respect to the inferential behaviour of each of the alternative formalisation candidates. Third, in spite of their *holistic* perspective on logic and language, the criteria they put forward merely assess the adequacy of formalisations of *single* sentences. The alternative formalisations of the same sentence are then put into some rigidly formalised sample arguments whose adequacy is taken for granted. Obviously, this seems at odds with the very task a theory of formalisation should primarily be concerned with since there is no reason for assuming that adequately formalising the other constituents of the argument is any less problematic than formalising sentence S.

3 Dialogical Criteria

3.1 A Dialogical Framework for Immanent Reasoning

As I am drawing on the most recent developments made by Rahman, McConaughey, Klev, and Clerbout (2018), I will briefly outline their current state of research. In their (2018) work *Immanent Reasoning or Equality in Action*, Rahman et al. develop what they call a dialogical framework *for immanent reasoning* by incorporating some features of Martin-Löf's *Constructive Type Theory* (for an introduction to CTT, see, e.g., Martin-Löf, 1984). Statements are now composed of a proposition and its local reason justifying the proposition, and thus have the form $X\ p : A$, which means that the player X has a (local) reason p for stating the proposition A (Rahman et al., 2018, p. 117). As a consequence, dialogical games for *immanent reasoning* – where *local reasons* are given and asked for – require some adjustments of the two basic sets of rules of dialogical logic, the *Particle Rules* and the *Structural Rules*. While the former prescribe how to appropriately attack and defend a proposition containing a specific logical constant, the latter govern the general development of a dialogical play (for an introduction to the basic notions of dialogical logic, see Rahman et al., 2018, chap. 3). The *Particle*

Rules now contain three different subsets of rules for each logical connective: the *Formation Rules*, the *Synthesis Rules*, and the *Analysis Rules* (for an informal introduction to these three types of rules, see Rahman et al., 2018, pp. 117–123). For example, when attacking an *existential* quantification in the course of a dialogical play, then, according to the respective *Analysis Rule* (see Rahman et al., 2018, p. 136, Table 7.4; Table 1 below is an extract of Table 7.4), the challenger can choose between asking for the left *or* for the right part of the quantification. It is then the *defender* who has the choice for the 'element in A' he wants to state in response. But when attacking an universal quantification, the challenging move consists in stating its left part and thereby demanding of the defender to state the requested proposition.

Table 1: *Analysis Rules for local reasons*

	Move	Challenge	Defence
Existential Q	X p: $(\exists x : A)\, B(x)$	$Y\ ?\ L^{\exists}$ or $Y\ ?\ R^{\exists}$	$X\ L^{\exists}(p)^{X}$: A (resp.) $X\ R^{\exists}(p)^{X}$: $B(L^{\exists}(p)^{X})$
Universal Q	X p: $(\forall x : A)\, B(x)$	$Y\ L^{\forall}(p)^{Y}$: A	$X\ R^{\forall}(p)^{X}$: $B(L^{\forall}(p)^{Y})$

3.2 Dialogical Principles of Reliability and Ambitiousness

I will start elaborating my dialogical criteria by proposing two modifications to Peregrin and Svoboda's principles. First, the formulation of the principles will be adapted to the dialogical approach by translating *logical validity* into *the existence of a P-winning strategy*. **P** is said to have a *winning strategy* for the thesis in question if for every possible choice of **O** to play, there is a way for **P** to win (for an introduction to the notion of *winning strategies*, and to the *strategy level* in general, see Rahman et al., 2018, chap. 5). Second, the new principles will implement, under some modification, Reinmuth's (n.d.) solution to the problem concerning holism. The following terminological remarks already indicate the shift from merely considering alternatively formalised *single* sentences to the evaluation of *entire sets* of formalisations. Note that I follow Peregrin and Svoboda's convention of using the term *formalisation* of S to denote the formula which is the result of a logical analysis of sentence S by means of a certain logical language. For a more detailed exposition of the relation between logical analysis, formalisation, formulas and instantiation, see Peregrin & Svoboda, 2017, pp. 62–64.

Dialogical Criteria of Adequate Formalisation

Terminology

S set of natural language sentences $\{S_1, \ldots, S_n\}$

A set of natural language arguments $\{A_1, \ldots, A_n\}$ over **S**

L logical system

ϕ set of formalisations $\{\phi_1, \ldots, \phi_n\}$ of **S** in **L** (i.e., sentence $S_1 \in$ **S** is a natural language instance of formalisation $\phi_1 \in \phi$)

Φ set of argument forms $\{\Phi_1, \ldots, \Phi_n\}$ over ϕ (i.e., argument $A_1 \in$ **A** is a natural language instance of argument form $\Phi_1 \in \Phi$)

With these preliminary remarks in mind, I propose the following reformulation of Peregrin and Svoboda's two criteria, which I call the *Dialogical Principle of Reliability* (d.REL) and the *Dialogical Principle of Ambitiousness* (d.AMB):

Definition 1 (d.REL) *ϕ counts as a set of adequate formalisations of S in L only if for every argument $A_k \in A$ it holds: if the dialogical game for the argument form $\Phi_k \in \Phi$ yields a P-winning strategy, then its natural language instance A_k is an intuitively correct argument.*

Definition 2 (d.AMB) *ϕ counts as a set of adequate formalisations of S in L only if for every argument $A_k \in A$ it holds: if the natural language instance A_k is an intuitively correct argument, then the dialogical game for its argument form $\Phi_k \in \Phi$ yields a P-winning strategy.*

It should be noted that I restrict the realm of application of the two criteria in the same way as Peregrin and Svoboda did. First, I only consider arguments that are *intuitively* (in-)correct, i.e., what the authors call *intuitively perspicuous* (2017, p. 64). The criteria can thus not be used to evaluate formalisations of highly complex arguments whose correctness cannot be intuitively assessed. Second, I adopt Peregrin and Svoboda's (2017, p. 65) *internal perspective* which presupposes that the chosen sample arguments fall into the *intended scope* of the system **L** and thereby takes into account the respective limitations of different logical systems. However, analogous to Reinmuth's approach, the two criteria are not confined to solely assessing the adequacy of formalisations of *single* sentences within a rigidly formalised frame. They thus take a *holisitic* perspective on the question of adequate formalisation: for a *set ϕ* of formalisations to count as adequate, there must be a bilateral correspondence between the intuitive (in-)correctness of the sample arguments in **A** and the (in-)validity of the respective argument forms in Φ. Since in the dialogical framework, logical validity coincides with the existence of a *P-winning strategy* (Rahman et al., 2018, p. 91), the two principles are formulated accordingly. So far, I have developed two *holistic* and *inferential* criteria of adequate formalisation. However, like in the original approach, the two criteria taken together merely tell us if some set counts as *a* set of adequate formalisations of a given set of sample arguments.

3.3 Dialogical Principle of Correspondence

Now *if* one is confronted with *more than one* set of adequate formalisations, it seems only natural to take a closer look at the syntactic variations between the alternative formalisations of the remaining sets. In what follows, I will offer an alternative to Peregrin and Svoboda's non-inferential auxiliary criteria and argue that one can maintain a purely inferentialist perspective by only *indirectly* taking into account the syntactic make-up of a formalisation. While the first two principles analyse how alternative sets of formalisations *behave in inferences*, the third principle will focus on the *role* syntactic variations between the remaining candidates *play* in this inferential behaviour.

This can be done by examining the varying distribution of *rights to challenge* and *duties to defend* that emerge during the dialogical plays on the alternative argument forms. One can then compare those *formal* dialogues with an *informal* argumentative strategy on the respective natural language argument which makes explicit the conditions to support the argument in terms of the *reasons* that are *given* and *asked for*. In short, the basic idea is to justify one alternative set of formalisations to be *more* adequate than another by comparing the degree of correspondence between the diverging development on the formal side with the pattern of *rights to ask for* and *duties to give reasons* on the informal side. Although the underlying concepts call for some explanation, I will first postulate the *Principle* (d.CRS).

Definition 3 (d.CRS) *ϕ counts (compared to ψ) as the set of more adequate formalisations of S in L if the distribution of rights to challenge and duties to defend emerging from the dialogical play for the argument form $\Phi_k \in \Phi$ corresponds more closely to the pattern of rights to ask for and duties to give reasons of the argumentative strategy for the corresponding argument $A_k \in A$.*

Dialogical Criteria of Adequate Formalisation

As implied in the formulation of my principle, its application is made up of three successive operations. First, on the *formal* side, one must identify the differences between the dialogical plays for each of the remaining formalisation candidates in terms of their diverging distribution of *rights to challenge* and *duties to defend*. Second, on the *informal* side, a natural language argumentative strategy on the corresponding sample argument has to be developed. Third, in order to decide which of the alternative sets turns out to be *more* adequate, the *degree of correspondence* between the varying distribution of challenges and defences has to be compared with the pattern of *reasons given* and *asked for* in the informal argumentative strategy. For the remainder of this paper, I will elaborate on each of the three operations.

On the *formal* side, one of the most distinctive features of the dialogical framework is its *rule-based* approach to meaning. As we have already seen in the case of the universal and the existential quantifier (Table 1), the *Particle Rules* prescribe in a dynamic fashion how to legitimately challenge and defend a proposition containing the relevant logical connective. My procedure thus takes advantage of the fact that the meaning of the logical constants is determined by the rules for dialogical interaction. In light of these rules, we can observe that when comparing *alternative* formalisations of the *same* argument, a different choice of a logical constant drastically changes the way a statement can be challenged and defended. Hence, the *Dialogical Principle of Correspondence* is sensitive to the inferential role of syntactic variations which is *beyond* the scope of analysis of the first two criteria. Now if one conceives dialogical games as *formal* argumentative strategies, a corresponding natural language discourse can be understood as an *informal* argumentative strategy.

Comparing the degree of correspondence between formal and informal argumentative strategies thus requires some objective procedure for constructing a natural language discourse. In order to provide such a procedure, I will draw on Brandom's (1994, pp. 167–198) *model of assertional practice*. Analysing natural language argumentation by making explicit the underlying pattern of *reasons given* and *asked for* will then allow for a meaningful comparison with the dialogical plays on the alternative argument forms. According to Brandom, argumentative interaction, i.e., the *game of giving and asking for reasons*, can be understood as a dynamic process of attributing commitments and entitlements to oneself as well as to others. When person E asserts an argument as correct, he *commits* himself to the premises and claims that such a commitment *entitles* him to the conclusion. His interlocutor F will thus attribute to E this very constellation of commitments and entitlements, called the *deontic score* (Brandom, 1994, p. 182). In making an assertion, E is consequently assuming responsibility to *demonstrate* his entitlement while at the same time committing himself to the various *inferential consequences* that follow (Brandom, 1994, p. 173). Now if F wants to challenge E's entitlement to an assertion, i.e., raise an objection, she can do so by committing herself to an assertion *incompatible* with it. E, in return, is then obliged to demonstrate his entitlement to the conclusion by providing appropriate reasons justifying that F's incompatible assertion, e.g., a negation of the conclusion, is *not* compatible with commitment to the premises. Two statements are *incompatible* if they stand in an entitlement-precluding inferential relation to each other (Brandom, 1994, p. 178). If it turns out that the conclusion *and* its negation, i.e., two mutually *incompatible* claims such that commitment to the first precludes entitlement to the second, are both *compatible* with commitment to the premises, then E must renounce his initial claim of entitlement to the conclusion. In fact, such an argumentative scheme reflects the most intuitive development of any dialectical discourse between two agents: if E can prove that his argument resists F's objection, it counts as correct since he has successfully demonstrated his entitlement to the conclusion. In light of Brandom's model, I propose the following procedure, made up of four stages, for developing an argumentative strategy revealing the pattern of *duties to give* and *rights to ask for reasons* that emerge in the process of justifying one's entitlement to the conclusion. Note that the applicability of all of the inferential criteria presupposes judgements of the (in-)compatibility between formulas on the formal and between propositions on the informal side.

Table 2: *procedure for developing an argumentative strategy*

	F	E
I	attributing to E change in *deontic score*	*asserting the argument*: commitment to P1 commitment to P2 entitlement to C
II	*challenging E's assertion*: commitment to ¬C compatible with (P1 & P2)?	

Dialogical Criteria of Adequate Formalisation

III		attempt to demonstrate incompatibility between ¬C and (P1 & P2) III.1) relating ¬C & P1 III.2) relating ¬C* & P2
IV.a	*challenge successful*: commitment to ¬C *is* compatible with (P1 & P2)	*failed* demonstration of entitlement to C
IV.b	*challenge failed*: commitment to ¬C is *not* compatible with (P1 & P2)	*successful* demonstration of entitlement to C

So in *stage I*, E asserts the argument by committing himself to the correctness of the inference, i.e., he claims that commitment to the premises (*Assume P*) *entitles* him to commitment to the conclusion (*Thus C*). F subsequently challenges E's assertion (*stage II*) by making an assertion *incompatible* with the conclusion: she commits herself to its negation. In *stage III*, E then attempts to demonstrate that ¬C is incompatible with commitment to (P1 & P2). To do so, E first inferentially relates ¬C and P1 (*stage III.1*): "if [*assumption of* ¬C], then (P1 *does* or *does not* rule out that) [*assumption of P1*]". A rational judgement about the compatibility will then yield a new proposition ¬C* which integrates the assumptions of P1 as a result of the preceding operation. In stage *III.2*, E goes on and inferentially relates ¬C* and P2: "if [*assumption of* ¬C*], then (it *is* or *is not* possible that) [*assumption of P2*]". A rational judgement will now tell us if ¬C* and P2 are compatible or not. If they turn out to be compatible (*stage IV.a*), then F's challenge was successful: "Thus, it is *possible* that [*assumption of* ¬C]". However, if they are *not* compatible (*stage IV.b*), then F's challenge has failed: "Thus, it is *impossible* that [*assumption of* ¬C]". Now the way stages II and III are put into practice for a specific argument will reveal the implicit pattern of *reasons given* and *asked for* which are then to be compared with the dialogical plays on the alternative formalisations. Note that the *round* brackets in E's utterance of *stage III.1* and *III.2* need only be included if one deals with what Brandom (1994, p. 168) calls a *permissive* inferential relation between the propositions at stake, they can be ignored for *committive* ones. In case that the conclusion constitutes a generic proposition, then, as an *inferential consequence* of such a commitment, F has further the right to challenge the argument by choosing a member of the relevant class for which E is subsequently obliged to demonstrate that the generic attribution applies. If the argument in question is composed of more than two premises, the procedure can of course be extended accordingly.

After having expanded on the underlying concepts and presuppositions, the meaning of (d.CRS) should now be a bit more illuminating. It should be clear however that the principle is only able to identify the set of *more* adequate formalisations *relative* to the proposed natural language argumentative strategy and essentially depends on the way such a strategy has been developed. To sum up the results so far, just like in Peregrin and Svoboda's theory, the dialogical approach first requires that a valid argument form will not yield instances of intuitively incorrect natural language arguments and vice versa. Only afterwards, one turns to considerations about the formalisations' syntactic make-up. But rather than relying on *non*-inferential auxiliary criteria, the *Dialogical Principle of Correspondence* constitutes a fully-fledged inferential criterion. By making use of the distinctive advantages of dialogical logic, the principle is sensitive to even the slightest syntactic variations between the remaining formalisations candidates. In addition, the approach no longer evaluates formalisations of *single* sentences in a rigidly formalised frame, but takes into consideration *entire sets*.

4 Example

I will now illustrate how the *Principle of Correspondence*, and in particular the third operation, can be put into practice. The sentence **S** for which Peregrin and Svoboda discuss some alternative formalisations reads *No grey donkeys are lazy* (2017, p. 66). The following (A_1) is one of the sample arguments they consider. Among the alternative formalisations they take into account, Ψ is the only one which turns out to meet the requirements of their two inferential criteria (2017, p. 67).

sample argument A_1	**formalisation Ψ of S in A_1**
Every donkey is a herbivore	$\forall x(Dx \rightarrow Hx)$
No herbivore is lazy	$\neg \exists x(Hx \wedge Lx)$
No grey donkeys are lazy	$\neg \exists x(Gx \wedge Dx \wedge Lx)$

Dialogical Criteria of Adequate Formalisation

Hence, there is no doubt that my dialogical version of their two principles will also qualify this candidate as adequate. It just needs to be transcribed into a type-theoretical grammar so that it becomes viable in a context of dialogical games for immanent reasoning. Note that according to my new terminology, ψ no longer denotes the formalisation of a *single* sentence, but an *entire set* of formalisations: $\psi = \{\psi_1, \psi_2, \psi_3\}$. Ψ_1, which is the only element of set Ψ of argument forms, is thus composed of the elements in ψ. As set ψ has passed the tests of *Reliability* and *Ambitiousness*, I choose an alternative set ϕ whose formalisations are *logically* equivalent to those in ψ in order to guarantee that *both* sets will pass the first two principles and thus count as adequate.

formalisation $\Phi_1 \in \Phi$ **formalisation $\Psi_1 \in \Psi$**

$(\forall x : A)(Dx \to Hx)$	ϕ_1	$(\forall x : A)(Dx \to Hx)$	ψ_1
$(\forall x : A)(Hx \to \neg Lx)$	ϕ_2	$\neg(\exists x : A)(Hx \land Lx)$	ψ_2
$(\forall x : A)((Dx \land Gx) \to \neg Lx)$	ϕ_3	$\neg(\exists x : A)((Dx \land Gx) \land Lx)$	ψ_3

Now in order to identify the set of *more* adequate formalisations by means of (d.CRS), one must first take a closer look at the diverging distribution of challenges and defences. Since the example only serves the purpose of illustration, I confine the analysis to the syntactic variation the two formal dialogues are dealing with first. When comparing the *Analysis Rule* for universal with the one for existential quantifiers (Table 1), it becomes evident that the two alternative argument forms engage *different* rights to challenge and duties to defend (for all the *Synthesis* and *Analysis Rules*, see Rahman et al., 2018, p. 136; Table 7.3 & 7.4; for the *Structural Rules*, see Rahman et al., 2018, pp. 138–144).

Table 3: *extract of dialogical play on argument form Φ_1*

	Opponent			Proponent	
				! $((\forall x : A)(Dx \to Hx) \land (\forall x : A)(Hx \to \neg Lx)) \to (\forall x : A)((Dx \land Gx) \to \neg Lx)$	0
1	m:= 1			n:= 2	2
3	p : $(\forall x : A)(Dx \to Hx) \land (\forall x : A)(Hx \to \neg Lx)$	0		q : $(\forall x : A)((Dx \land Gx) \to \neg Lx)$	4

5	$L^\forall(q) : A$	4		$R^\forall(q) :$ $(Dq_1 \wedge Gq_1) \to \neg Lq_1$	8
7	$q_1 : A$		5	$?.../L^\forall(q)$	6
9	$?.../R^\forall(q)$	8		$q_2 : (Dq_1 \wedge Gq_1) \to \neg Lq_1$	10
	...			...	

P wins

In the dialogical play on argument form Φ_1 (Table 3), the *right to challenge* the universal quantification of *move 4* consists in stating its left part (*move 5*). After the instruction has been resolved, this amounts to *proposing* 'some element' (q_1) in A in *move 7*. **P**'s *duty to defend* (*moves 8* and *10*) is then to provide a reason justifying that for *this* element (q_1), the predicates within the scope of the universal quantification apply. The distribution of challenges and defences is strikingly different in the dialogical play on argument form Ψ_1 (Table 4). Here, the *right to challenge* the negated existential quantification of *move 4* first consists in attacking the negation in *move 5*. Since **O**'s challenge on the negation implies a distribution of the burden of proof, **P** can challenge the existential quantification from *move 8* onwards after having requested a resolution of the instruction in *move 6*. In *move 8*, he first asks for the *left*, and then, in *move 12*, for *right* part of the existential quantifier. In contrast to a legitimate attack on an universal quantification, the right to challenge the existential quantification is thus limited to *asking* one's adversary to choose an appropriate element ($q_{1.1}$ in *move 11*) for which it must then be demonstrated that the predication applies.

Table 4: extract of dialogical play on argument form Ψ_1

	Opponent			**Proponent**	
				$!\,((\forall x : A)\,(Dx \to Hx)$ $\wedge \neg(\exists x : A)\,(Hx \wedge Lx)) \to$ $\neg(\exists x : A)((Dx \wedge Gx) \wedge Lx)$	0
1	m:= 1			n:= 2	2
3	$p : (\forall x : A)\,(Dx \to Hx)$ $\wedge \neg(\exists x : A)\,(Hx \wedge Lx)$	0		$q :$ $\neg(\exists x : A)((Dx \wedge Gx) \wedge Lx)$	4
5	$L^\neg(q) :$ $(\exists x : A)((Dx \wedge Gx) \wedge Lx)$	4		—	
7	$q_1 :$ $(\exists x : A)((Dx \wedge Gx) \wedge Lx)$		5	$?.../L^\neg(q)$	6

Dialogical Criteria of Adequate Formalisation

9	$L^{\exists}(q_1) : A$		7	$? L^{\exists}$	8
11	$q_{1.1} : A$		9	$?.../L^{\exists}(q_1)$	10
13	$R^{\exists}(q_1) :$ $(Dq_{1.1} \wedge Gq_{1.1}) \wedge Lq_{1.1}$		7	$? R^{\exists}$	12
15	$q_{1.2} :$ $(Dq_{1.1} \wedge Gq_{1.1}) \wedge Lq_{1.1}$		13	$?.../R^{\exists}(q_1)$	14
	...			...	

P wins

Let's suppose that the procedure (Table 2) for developing an informal argumentative strategy for A_1 yields the following dialogue:

	E(1):	Assume that every donkey is a herbivore [*commitment to P1*]
(I)	E(2):	Assume that no herbivore is lazy [*commitment to P2*]
	E(3):	Thus, no grey donkeys are lazy [*claiming entitlement to C*]
	F(4):	Assume that (at least some) grey donkeys are lazy: assume
(II)		Batu is a grey donkey [*commitment to incompatible claim precludes entitlement to C*]
	E(5):	If Batu is a grey donkey, then he is a donkey
(III)	E(6):	If Batu is a donkey, then he is a herbivore [*III.1*]
	E(7):	If Batu is a herbivore, then he is not lazy [*III.2*]
(IV.b)	F(8):	Thus, it is impossible that (at least some) grey donkeys are lazy

As can be observed, E's assertion of the argument (*stage I*) entails F's *right to ask for reasons* justifying that for *any* grey donkey, it is impossible to be lazy (*stage II*). E's *duty to give reasons* in *stage III* then consists in demonstrating that, based on his commitment to the premises, it is not possible for such a randomly chosen grey donkey to be lazy. In other words, as an *inferential consequence* of his entitlement claim, he must demonstrate that the generic proposition of the conclusion necessarily applies to the grey donkey chosen by F. He does so by inferentially relating $\neg C$ with P1 and thereby providing a reason for the grey donkey Batu being a herbivore (*III.1*). He then (*III.2*) provides a further reason that the herbivore Batu cannot be lazy by relating $\neg C^*$ with P2. Hence, the generic attribution of the conclusion necessarily applies to the class member F has picked out. As a result (*stage IV.b*), he has successfully demonstrated that his commitment to the premises entitles him to commitment to the conclusion.

Table 5: *final comparison (operation 3)*

	formal dialogue on Φ_1	formal dialogue on Ψ_1	informal dialogue	
right to challenge	*proposing* 'some element' in A: **O**(5): $L^\forall(q) : A$	*asking* to bring forward 'some element' in A: **P**(8): ? $L^\exists$ **P**(12): ? $R^\exists$	*proposing* 'some member' of class *grey donkeys*: **F**(4): assume Batu is a grey donkey	*right to ask for reasons*
duty to defend	justifying that for *this* element (chosen by **O**), the predicates apply: **P**(8): $R^\forall(q) : (Dq_1 \wedge Gq_1) \rightarrow \neg Lq_1$	justifying that for *this* element (chosen by **P**), the predicates apply: **O**(9): $L^\exists(q_1) : A$ **O**(13): $R^\exists(q_1) : (Dq_{1.1} \wedge Gq_{1.1}) \wedge Lq_{1.1}$	justifying that for *this* member (chosen by **F**), the generic attribution applies: **E**(5-7): If Batu is a grey donkey, (...) then he is not lazy	*duty to give reasons*

Even though both formalisations have passed the tests of *Reliability* and *Ambitiousness*, i.e., count as adequate, the application of (d.CRS) now suggests that, *relative* to the proposed informal argumentative strategy, ϕ might turn out to be the set of *more* adequate formalisations (Table 5). Just like on the *informal* side, the *formal* dialogue on Φ_1 entails **O**'s *right to challenge* **P**'s universal quantification by proposing some element q_1 of Type A. Thus, at least with respect to the first syntactic variation, the distribution of *challenges* and *defences* of the formal play on Φ_1 corresponds in its structure *more closely* to the pattern of *reasons given* and *asked for* in the informal argumentative dialogue on A_1. The procedure of (d.CRS) must of course be carried out for all the syntactic variations in order to make a conclusive judgement.

5 Conclusion

Although I have restored the inferential and holistic features which are at the bottom of Peregrin and Svoboda's project, there are at least two concerns which make my approach not yet entirely satisfactory. First, the proposed procedure for developing an informal argumentative strategy does not, in its present form, capture every possible type of argument. Rather than allowing for a strict application, it must be adapted to the individual cases one is dealing with. Hence, a certain degree of ambiguity with respect to the outcome of applying (d.CRS) seems inevitable. Second, the procedure, and in particular *stage III*, relies on the possibility that human reason alone allows to decide whether two propositions are compatible or not. Such a judgement might however exceed the limits of our rational capacities when being confronted with more complex cases. Hence, the dialogical criteria I have developed are not yet an entirely flawless tool for adequately formalising natural language arguments. But having illustrated in some detail one possible way of using the dialogical toolbox in this area—an idea which has so far been neglected by both formalisation and dialogical theorists—I hope to have provided an interesting and somewhat new perspective on the question of adequate formalisation. In order to make the approach more resistant to objections, the procedure for constructing the informal argumentative strategy should be revised such that one can follow it more strictly and apply it more generally. It might also be interesting for future research to try taking into consideration the *Formation Rules* for logical constants (Rahman et al., 2018, pp. 132–135). Linking adequacy with *Formation-Plays* could thus yield an alternative, perhaps complementary, way of applying the dialogical approach to the theory of formalisation.

References

Brandom, R. (1994). *Making it Explicit. Reasoning, Representing, and Discoursive Commitment.* Cambridge: Havard University Press.

Martin-Löf, P. (1984). *Intuitionistic Type Theory. Notes by Giovanni Sambin of a Series of Lectures given in Padua, June 1980.* Naples: Bibliopolis.

Peregrin, J., & Svoboda, V. (2017). *Reflective Equilibrium and the Principles of Logical Analysis. Understanding the Laws of Logic.* New York: Taylor and Francis.

Rahman, S., McConaughey, Z., Klev, A., & Clerbout, N. (2018). *Immanent Reasoning or Equality in Action. A Plaidoyer for the Play Level.* Cham: Springer International Publishing AG.

Reinmuth, F. (n.d.). *Holistic Inferential Criteria of Adequate Formalization.* (To appear in *Dialectica*)

Simon Brausch
Humboldt University of Berlin, Department of Philosophy
Germany
E-mail: s.d.brausch@googlemail.com

Relative Interpretations and Substitutional Definitions of Logical Truth and Consequence

MIRKO ENGLER

Abstract: This paper proposes substitutional definitions of logical truth and consequence in terms of relative interpretations that are extensionally equivalent to the model-theoretic definitions for any relational first-order language. Our philosophical motivation to consider substitutional definitions is based on the hope to simplify the meta-theory of logical consequence. We discuss to what extent our definitions can contribute to that.

Keywords: substitutional definitions of logical consequence, Quine, metatheory of logical consequence

1 Introduction

This paper investigates the applicability of relative interpretations in a substitutional account of logical truth and consequence. We introduce definitions of logical truth and consequence in terms of relative interpretations and demonstrate that they are extensionally equivalent to the model-theoretic definitions as developed in (Tarski & Vaught, 1956) for any first-order language (including equality). The benefit of such a definition is that it could be given in a meta-theoretic framework that only requires arithmetic as axiomatized by PA. Furthermore, we investigate how intensional constraints on logical truth and consequence force us to extend our framework to an arithmetical meta-theory that itself interprets set-theory. We will argue that such an arithmetical framework still might be in favor over a set-theoretical one.

The basic idea behind our definition is both to generate and evaluate substitution instances of a sentence φ by relative interpretations. A relative interpretation rests on a function f that translates all formulae φ of a language L into formulae $f(\varphi)$ of a language L' by mapping the primitive predicates P of φ to formulae ψ_P of L' while preserving the logical structure of φ

and relativizing its quantifiers by an L'-definable formula. In that sense, the application of translation functions boils down to the uniform substitution of predicates by complex formulae, which was the method used by Quine (1970) to give a substitutional definition of logical truth and consequence.

Substitutional definitions of logical truth and consequence play an important role in the history of logic. As it appears, a substitutional criterion has been traditionally the preferred choice for an analysis of what a logical truth is. However, since the seminal work of Tarski (1936) on the concept of logical truth and consequence, substitutional definitions were considered obsolete in the 20th Century by most of the logicians. An important exception to that is, as mentioned, Quine, who was interested in avoiding the ontological commitments of set-theory, which is the required meta-theory for a model-theoretic definition.[1] Still, it might seem perplexing to any modern logician that the most basic notion in logic, the notion of logical consequence, needs to be explained in the opulent realm of set-theory. At least, let it be our motivation to take up Quine's project to give a substitutional definition of logical consequence in order to avoid the use of set-theory as far as possible.

Another motivation for considering a substitutional definition of logical truth and consequence instead of a model-theoretic one stems from the alleged problem of truth preservation in a model-theoretic setting. In this context "truth preservation" means that from the logical truth of a sentence one can infer its truth simpliciter. It has been claimed that a model-theoretic setting fails to give a so-called "intended" interpretation of a language[2] by the notion of truth-in-a-model. Thus, an adequate explanation of "truth simpliciter" is missing in a model-theoretic setting. A discussion of that issue and a substitutional definition that trivially seems to fulfill truth preservation has been recently given by Halbach (2018). In the present paper, however, we will only focus on the aspect of avoiding a set-theoretic meta-theory for logical consequence.

The paper proceeds as follows: Section 2 gives a short outline of Tarski's reasons to reject substitutional definitions of logical truth and consequence as well as Quine's idea of how to improve substitutional definitions. By pointing out the limits of his definition in (Quine, 1970), we also try to motivate our framework of relative interpretations. Section 3 introduces

---

[1] For an outline of the development of Quine's conception of logical truth and consequence, see (Wagner, 2019).

[2] One example mentioned by Halbach (2018) is that in a model-theoretic framework the universal quantifier has to be evaluated as a set. For a set-theoretic language, however, this seems to be inadequate since the collection of all sets is a class and not a set.

a definition of logical truth and consequence and proves its extensional adequacy. The framework of this definition can be seen to only require PA if the antecedent-set of a logical consequence is assumed to be recursively enumerable. Nevertheless, we have to point out that this definition might be considered inadequate from an intensional perspective as compactness is used to define consequence and a corresponding notion of satisfaction is not available. Subsequently, Section 4 considers another definition that is extensionally adequate but also meets the mentioned intensional constraints. However, the required meta-theory for that definition itself interprets set-theory. Section 5 discusses in how far these results still can be considered as a method of avoiding set-theory in the meta-theory of logical consequence.

2 Substitutional Definitions of Logical Consequence: Tarski and Quine

A definition of logical truth and consequence on the basis of the substitution of syntactic particles in sentences is apparently close to an informal characterization of logical truth and consequence. Traditionally, the understanding of a logical consequence is reflected in criteria of truth preservation in virtue of form, wherein it is suggested to explain "form" in terms of substitution. As is well known, even Tarski considered a substitutional criterion for an adequate analysis of logical consequence in (Tarski, 1936). There he states a condition (F), which aims to characterize a logical consequence in the following way:

> (F) If, in the sentences of the class K and in the sentence X, the constants—apart from the purely logical constants—are replaced by any other constants (like signs are being everywhere replaced by like signs) and if we denote the class of sentences thus obtained from K by K', and the sentence obtained from X by X', then the sentence X' must be true provided only that all sentences of the class K' are true.[3]

Tarski considered condition (F) to be necessary for any definition of logical consequence. However, he proceeds with the claim:

> The condition (F) could be regarded as sufficient for the sentence X to follow from the class K only if the designations of all possible objects occurred in the language in question. This assumption, however, is fictitious and can never be realized.

---
[3]Translation of (Tarski, 1936) in (Tarski, 1956, pp. 409–420) by J.H.Woodger.

Here, Tarski indicates that a substitutional characterization of logical consequence cannot deal with a language that has not enough non-logical expressions in order to make a true but non-logical true conclusion false by a substitution. Whereas in a model-theoretic setting it would be possible to vary the evaluation of the non-logical constants over the domain of a certain model to eliminate this problem. The assumption that no substitutional method can compete with a set-theoretic method might seem plausible. Notably, the antinomies of Russell and Grelling show that there are formulae that don't determine a set as well as sets that cannot be determined by any formula. Nevertheless, it will turn out as a misconception that a substitutional method cannot be adequate for any language.

Quine's improvement of the substitutional method rests on the idea of a uniform substitution of predicates by complex formulae of a language, which provides an increased range of possible substitution instances. If a first-order relational language L (without equality) is strong enough[4] to express elementary arithmetic, then - Quine claimed - such a substitutional notion of logical consequence can be extensionally equivalent to the model-theoretic notion. An accessible presentation of that idea can be found in (Ebbs & Goldfarb, 2018).

Nonetheless, Quine's remarkable achievement is evidently still of minor generality compared to the model-theoretic treatment as it leaves out languages that cannot express elementary arithmetic or languages that consider equality as a logical constant. A crucial hurdle for any substitutional definition of logical truth and consequence is the treatment of the equality relation. For any language L that considers "=" as a logical constant we have (contingent) sentences without non-logical symbols (like "$\exists v_1 \exists v_2 (v_1 \neq v_2)$"). Consequentially, no symbol can be substituted since all symbols are logical ones. So the difficult question would be how to generate substitution instances that can come out false in the language L—especially if L is required to express elementary arithmetic.

Another problem for Quine but in general for any substitutional account that considers the preservation of truth under substitution arises from the notion of truth itself. Any definition of an adequate notion of truth for a language L that expresses elementary arithmetic cannot be given in the arithmetical language L. Any language L^+ that is capable of expressing the notion of truth for L must have expressive power exceeding that of L.

---

[4]This means that the language contains, inter alia, predicates that can be evaluated in a structure as the basic arithmetic relations of being the number zero, being the successor of a number, the sum of two numbers and the product of two numbers.

For that reason, a substitutional definition of logical consequence, where the notion of truth for L is defined in L^+, would not be applicable to L^+. As is well known, Tarski's original account of logical consequence in (Tarski, 1936) was also confronted with that problem, which led him to the model-theoretic framework of (Tarski & Vaught, 1956). While the definition of logical consequence is given in the language of set-theory, the definition is applicable to the language of set-theory itself. If "truth" is not taken as primitive, then the challenge for any substitutional definition will be not to fall back to the use of model-theory while preserving its universal applicability. Conclusively, we will critically have to review to what extent our definition is able to master this challenge.

3 Interpretations and Logical Truth and Consequence

The first improvement of our substitutional definition is based on the substitution of the predicate symbols of a language L by complex syntactic particles from *another* language. In addition to Quine's improvement of the substitution of predicates by *complex* formulae, we also allow these formulae to be taken (uniformly) from another language. We choose the arithmetical language of $L[\text{PA}]$. Thereby, we again extend the range of the available substitution instances for a language to a sufficient level and bypass the problem of excluding languages that are not rich enough to express arithmetic from the definition (in the sense explained above).

A convenient way of generating these substitution instances for any language is to use relative translation functions as they first appeared in (Tarski, Mostowski, & Robinson, 1953) to define relative interpretations.[5] To keep things simple, we only want to consider relational languages to be translated. Nevertheless, all results also apply to languages with constant and function symbols due to their relational eliminability and of course, the translating language may always contain constant and function symbols.

---

[5] A relative interpretation of a theory S in a theory T is a relative translation function f (as explained in Section 1) from $L[S]$ to $L[T]$ s.t. T proves all the translated theorems of S. For that, we shortly write $S \prec_f T$ and $S \prec T$ if there is a translation function f s.t. $S \prec_f T$.

Mirko Engler

Definition 1 (Substitution Function) *Let L be a relational language of first-order logic with equality. Then a* substitution function *f for L is given by a function $I : L \to L[\mathrm{PA}]$ assigning a formula of $L[\mathrm{PA}]$ in n variables to every n-ary predicate symbol of L and a formula $\delta(v)$ of $L[\mathrm{PA}]$ with exactly one free variable v such that:*

1. $f(v_{i_n} = v_{i_m}) \doteq (v_{i_n} = v_{i_m})$[6]

2. $f(Pv_{i_1}...v_{i_n}) \doteq I(P)(v_{i_1}...v_{i_n})$[7]

3. $f(\neg\varphi) \doteq \neg f(\varphi)$ and $f(\varphi \to \psi) \doteq f(\varphi) \to f(\psi)$

4. $f(\forall v_i \varphi) \doteq \forall v_i(\delta(v_i) \to f(\varphi))$

5. $\mathrm{PA} \vdash \exists v \delta(v)$

A substitution function is simply a relative translation function where the translating language is $L[\mathrm{PA}]$ and the formula δ, which defines in $L[\mathrm{PA}]$ the domain for evaluating the quantifier, can be proven to be non-empty in PA (condition 5).[8] This condition ensures that a substitution function preserves logical truth in PA, which is the subject of Lemma 1 below.

The advantage of a translation function that can relativize quantifiers comes into effect when we require an adequate treatment of "=". As mentioned, a substitutional definition of logical truth and consequence for a language including "=" as a logical constant is confronted with the problem of defining substitution instances for (contingent) sentences without non-logical symbols (like "$\neg \forall v_1 \forall v_2 (v_1 = v_2)$"). Since we cannot substitute any symbol in those sentences in order to generate a substitution instance that is unsatisfiable, the relativization of quantifiers offers a possibility to deal with that problem. For instance, a substitution function f where $\delta(v)$ equals "$v = \overline{0}$"[9] translates the sentence "$\neg \forall v_1 \forall v_2 (v_1 = v_2)$" to "$\neg \forall v_1 (v_1 = \overline{0} \to \forall v_2 (v_2 = \overline{0} \to v_1 = v_2))$". Trivially, $\mathrm{PA} \vdash \exists v \delta(v)$,

---

[6]The symbol "$\doteq$" denotes equality in the meta-language.

[7]The substitution of the variables in $I(P)$ by $v_{i_1},...,v_{i_n}$ may cause a collision of variables, which means that variables which were free in the original formula are bound by quantifiers in $I(P)$. In such a case we rename the bounded variables of $I(P)$ by $v_{i_{m+1}},...,v_{i_{m+k}}$, where m is the maximum of the indices of the variables $v_{i_1},...,v_{i_n}$.

[8]Sometimes a similar condition can be found in definitions of relative interpretability, namely that the interpreting theory should prove the relativization δ of a translation f to be non-empty. However, if f is a relative interpretation, then the interpreting theory trivially proves "$\exists v \delta(v)$" for it has to interpret "$\exists v v = v$" as a theorem of logic.

[9]The symbol "$\overline{0}$" is an individual constant of $L[\mathrm{PA}]$ which is meant to denote 0.

Interpretations and Logical Consequence

so we gave a substitution function f such that the substitution instance of "$\neg \forall v_1 \forall v_2 (v_1 = v_2)$" is not provable in arithmetic. Moreover, its negation will be provable and we have got sufficient means to classify the sentence "$\neg \forall v_1 \forall v_2 (v_1 = v_2)$" as not logically true by our substitutional method.

The second improvement of our definition is based on using relative interpretations in order to omit the problem of explaining the notion of truth for any language. Instead of considering the invariance of truth under substitution, we consider the invariance of interpretability in PA under substitution for the classification of a logical truth. That this turns out to be sufficient is due to Lemma 2 below, which roughly states that for a consistent set of sentences Γ one can define in PA a model for Γ by assuming the formal consistency of Γ. Usually, one considers theories as the objects that are interpreted. However, we can extend the range of relative interpretations to sets of sentences if we take care of the fact that they are not always deductively closed. To define a logical consequence in this framework, we utilize that consequence can be classified in terms of logical truth due to the compactness of logical consequence. In that respect, our definition also resembles Quine's definition in (Quine, 1970).

Definition 2 (Logical Truth and Consequence) *Let φ be any sentence of a relational language L of first-order logic with equality, Γ a set of L-sentences;*

1. $LTr(\varphi) :\Leftrightarrow \mathbb{V} f(subst(f) \Rightarrow \text{PA} \vdash f(\varphi))$[10]

2. $LConseq(\Gamma, \varphi) :\Leftrightarrow \exists \Gamma' (\Gamma' \text{ is finite} \mathbin{\mathbb{\wedge}} \Gamma' \subseteq \Gamma \mathbin{\mathbb{\wedge}} LTr(\bigwedge \Gamma' \to \varphi))$

Lemma 1 (Correctness) *Let φ be any sentence of a relational language L of first-order logic with equality, then $\models \varphi \Rightarrow \mathbb{V} f(subst(f) \Rightarrow \text{PA} \vdash f(\varphi))$.*

Proof. Assume $\models \varphi$ and let f be a substitution function. Define a monotone operator Π on sets of L-sentences considering a usual list of logical axioms and rules of inference for first-order logic such that the logical truths are the smallest set which is closed under Π, i.e., $\models \varphi$ is equivalent to $\mathbb{V} X(\Pi(X) \Rightarrow \varphi \in X)$ and it holds that $\mathbb{V} X(\Pi(X) \Rightarrow (\models \varphi \Rightarrow \varphi \in X))$. Let $Y := \{\varphi \in Sent_L \mid \text{PA} \vdash f(\varphi)\}$. It can be easily shown that $\Pi(Y)$ holds and so $\text{PA} \vdash f(\varphi)$. □

---

[10] We use double-lined logical symbols like "$\mathbb{V}$" for the logical constant of the meta-language and "$subst(f)$" as a shorthand for "f is a substitution function".

Mirko Engler

Lemma 2 (Formalized Completeness) *Let T be a consistent extension of PA in $L[\text{PA}]$, Γ be a set of sentences of a relational language L of first-order logic with equality s.t. its deductive closure is numerated[11] in T by an $L[\text{PA}]$-formula γ, then $\Gamma \prec T + \text{Con}_\gamma$.*[12]

Corollary 1 *Let Γ be a consistent set of sentences of a relational language L of first-order logic with equality s.t. its deductive closure is numerated in PA by an $L[\text{PA}]$-formula γ, then $\exists f(subst(f) \wedge \Gamma \prec_f \text{PA} + \text{Con}_\gamma)$.*

Proof. By Lemma 2, it immediately follows that there is a relative translation function $f : L \to L[\text{PA}]$ such that $\text{PA} \vdash f(\psi)$ for all sentences ψ s.t. $\Gamma \vdash \psi$. The proof of Lemma 2 which can be found in (Lindström, 1997, §6) shows that already $\text{PA} \vdash \exists v \delta(v)$, where δ is the relativization of a relative translation that interprets Γ in $\text{PA} + \text{Con}_\gamma$. □

Proposition 1 *Let φ be any sentence of a relational language L of first-order logic with equality and Γ a set of L-sentences, then;*

1. $\models \varphi \Leftrightarrow LTr(\varphi)$

2. $\Gamma \models \varphi \Leftrightarrow LConseq(\Gamma, \varphi)$

Proof. (1.) : Follows from (2.) for $\Gamma = \emptyset$.

(2. $\Rightarrow$) : Assume that $\Gamma \models \varphi$. Therefore, there is a finite subset Γ' of Γ such that $\models \bigwedge \Gamma' \to \varphi$. By Lemma 1, it follows that $\forall f(subst(f) \Rightarrow \text{PA} \vdash f(\bigwedge \Gamma' \to \varphi))$. So $LConseq(\Gamma, \varphi)$.

(2. $\Leftarrow$) : Assume there is a finite subset Γ' of Γ s.t. $\forall f(subst(f) \Rightarrow \text{PA} \vdash f(\bigwedge \Gamma' \to \varphi))$ and $\Gamma \not\models \varphi$. From the latter assumption it follows that $\Gamma \cup \{\neg\varphi\}$ is consistent. Furthermore, $\Gamma' \cup \{\neg\varphi\}$ is consistent. Since $\Gamma' \cup \{\neg\varphi\}$ is finite, its deductive closure can be numerated in PA by a formula γ^* in Σ_1^0. By Corollary 1, there is a substitution function f s.t. $\text{PA} + \text{Con}_{\gamma^*} \vdash f(\bigwedge \Gamma' \wedge \neg\varphi))$. By definition, $f(\bigwedge \Gamma' \wedge \neg\varphi)$ is equivalent to $\neg f(\bigwedge \Gamma' \to \varphi)$. By assumption, however, $\text{PA} + \text{Con}_{\gamma^*} \vdash f(\bigwedge \Gamma' \to \varphi)$, which means that $\text{PA} + \text{Con}_{\gamma^*}$ is inconsistent. As we assume PA to be consistent, it holds that $\text{PA} \vdash \neg \text{Con}_{\gamma^*}$. This is a contradiction as $\Gamma' \cup \{\neg\varphi\}$ is consistent and PA is assumed to be Σ_1^0-sound.[13] So $\Gamma \models \varphi$. □

---

[11]An $L[\text{PA}]$-formula γ numerates a set Γ in T iff $\varphi \in \Gamma \Leftrightarrow T \vdash \gamma(\overline{\ulcorner\varphi\urcorner})$ for all φ, where $\overline{\ulcorner\varphi\urcorner}$ denotes the gödelnumeral of φ. γ bi-numerates Γ in T if additionally, it holds that $\varphi \notin \Gamma \Leftrightarrow T \vdash \neg\gamma(\overline{\ulcorner\varphi\urcorner})$ for all φ.

[12]Originally, see (Feferman, 1960, Theorem 6.2).

[13]This means $\text{PA} \vdash \varphi$ implies $\mathfrak{N} \models \varphi$ if $\varphi \in \Sigma_1^0$. The Σ_1^0-soundness of PA is equivalent to its 1-consistency, a purely syntactical notion, which is ω-consistency restricted to p.r. formulas. For more details, see, e.g., (Smorynski, 1977, §4).

Interpretations and Logical Consequence

Proposition 1 shows that it is possible to give a substitutional definition of logical truth and consequence which is extensionally equivalent to the model-theoretic definition and only seems to require arithmetic as a meta-theory. Apart from that we will have to relativize this claim in the last section, it can be already pointed out that the definition still has some obvious flaws on the intensional level. For instance, if we expect the notion of logical truth defined in terms of relative interpretations to behave in the same way as its counterpart defined in terms of models, this might include the request that giving a structure which satisfies $\neg\varphi$ in order to show that φ is not a logical truth corresponds to giving a substitution function f such that $PA \vdash f(\neg\varphi)$. This is to request that $\neg Ltr(\varphi) \Leftrightarrow \exists f(subst(f) \wedge PA \vdash f(\neg\varphi))$. As a reason for that, one could think of this equivalence as a feature of logical truth that ensures its semantical character. However, by definition of $\neg Ltr(\varphi)$, we only know that there is a substitution function f such that $PA \not\vdash f(\varphi)$. But since PA is incomplete, this cannot be equivalent to $PA \vdash f(\neg\varphi)$ for any sentence φ of $L[PA]$.[14]

This fact corresponds to the observation that we cannot have a definition of the notion of satisfaction in terms of relative interpretations in PA (or in any consistent, recursively enumerable extension of PA). Any adequate notion of satisfaction would have to fulfill that φ being not a logical consequence of Γ is equivalent to the satisfiability of $\Gamma + \neg\varphi$. For our substitutional definition of logical consequence, this also fails for reasons of incompleteness. The conclusion we can draw from this is that an adequate substitutional definition of logical truth and consequence should come with a corresponding notion of satisfaction.

Another point, which has been originally made by Boolos (1975), is that a substitutional definition of logical consequence is of minor generality compared to the model-theoretic one if the compactness of the consequence relation is built-in to its definition—like in our case. Moreover, it may turn out that if the consequence relation is defined without built-in compactness, compactness doesn't hold for a so defined consequence relation.[15] Both issues, that of a corresponding notion of satisfaction and that of compactness will be taken care of in the following section.

---

[14] For a counterexample, consider a sentence φ that expresses in $L[PA]$ the inconsistency of PA. Trivially, φ is not a logical truth and $\neg\varphi$, which expresses the consistency of PA, is not interpretable in PA (see Feferman, 1960).

[15] See (Eder, 2016) for an outline of that discussion.

4 Interpretations, Satisfaction and Compactness

To meet the requirements which were brought up in the previous section, we restrict the following substitutional definition of satisfaction and logical consequence to arithmetically definable sets[16] of assumptions and extend our arithmetical background by the set $\text{Tr}(\Pi^0_{n+1})$ for a given n, which characterizes in $L[\text{PA}]$ the set of $L[\text{PA}]$-sentences that are of complexity Π^0_{n+1} and true in the standard model. Thereby, we have got an arithmetical theory which proves the consistency of any consistent set of sentences that is arithmetically definable in Π^0_n. It is important to note that the set $\text{Tr}(\Pi^0_{n+1})$ can be defined purely syntactically and we don't have to refer to the model-theoretic definition of satisfaction.[17]

Definition 3 *Let φ be a sentence of a relational language L of first-order logic with equality and Γ a set of L-sentences s.t. the deductive closure of Γ is arithmetically definable by an $L[\text{PA}]$-formula in Π^0_n (if $n > 0$, Σ^0_1 otherwise), then*

1. $Sat^n(\Gamma) :\Leftrightarrow \exists f(subst(f) \wedge \Gamma \prec_f \text{PA} + \text{Tr}(\Pi^0_{n+1}))$

2. $LConseq^n(\Gamma, \varphi) :\Leftrightarrow \forall f(subst(f) \Rightarrow (\Gamma \prec_f \text{PA} + \text{Tr}(\Pi^0_{n+1}) \Rightarrow \varphi \prec_f \text{PA} + \text{Tr}(\Pi^0_{n+1})))$

3. $LTr^*(\varphi) :\Leftrightarrow LConseq^0(\emptyset, \varphi)$

Corollary 2 *Let Γ be a set of sentences of a relational language L of first-order logic with equality s.t. the deductive closure of Γ is arithmetically definable by an $L[\text{PA}]$-formula in Π^0_n (if $n > 0$, Σ^0_1 otherwise). If Γ is consistent, then $\exists f(subst(f) \wedge \Gamma \prec_f \text{PA} + \text{Tr}(\Pi^0_{n+1}))$.*

The Corollary follows from Lemma 2, the observation that $\text{PA} \vdash \exists v \delta(v)$, where δ is the relativization of a relative translation that interprets Γ in $\text{PA} + \text{Tr}(\Pi^0_{n+1})$, and the fact that any arithmetically definable set Γ in Π^0_n can be numerated in $\text{PA} + \text{Tr}(\Pi^0_{n+1})$, which is the subject of the following Lemma.

Lemma 3 *Let Γ be a set of sentences of a relational language L of first-order logic with equality s.t. Γ is arithmetically definable by an $L[\text{PA}]$-formula γ in Π^0_n, then γ bi-numerates Γ in $\text{PA} + \text{Tr}(\Pi^0_{n+1})$.*

---

[16] An $L[\text{PA}]$-formula γ arithmetically defines a set of sentences Γ iff $\varphi \in \Gamma \Leftrightarrow \mathfrak{N} \models \gamma(\ulcorner \varphi \urcorner)$ for all φ, where $\mathfrak{N}$ denotes the standard model for $L[\text{PA}]$.

[17] For a definition, see, e.g., (Kaye, 1991, § 9.3).

Proof. Assume there is a γ in Π_n^0 s.t. $\psi \in \Gamma \Leftrightarrow \mathfrak{N} \models \gamma(\ulcorner\psi\urcorner)$. Since $\gamma(\ulcorner\psi\urcorner)$ is in Π_n^0, if $\psi \in \Gamma$, then $\mathrm{PA} + \mathrm{Tr}(\Pi_{n+1}^0) \vdash \gamma(\ulcorner\psi\urcorner)$. Assume now $\mathrm{PA} + \mathrm{Tr}(\Pi_{n+1}^0) \vdash \gamma(\ulcorner\psi\urcorner)$ but $\psi \notin \Gamma$, which means that $\mathfrak{N} \models \neg\gamma(\ulcorner\psi\urcorner)$. Since $\neg\gamma(\ulcorner\psi\urcorner)$ is in Σ_n^0, we would have $\mathrm{PA} + \mathrm{Tr}(\Pi_{n+1}^0) \vdash \neg\gamma(\ulcorner\psi\urcorner)$, which leads to a contradiction since we assume that $\mathrm{PA} + \mathrm{Tr}(\Pi_{n+1}^0)$ is consistent. Analogously, it follows that γ is a bi-numeration. □

The compactness of Sat^n follows from the following Lemma, which generalizes Orey's Compactness Theorem in (Orey, 1961, Theorem 3.1) to arithmetically definable sets. As he shows in (Orey, 1961, Theorem 3.2), his Compactness Theorem fails for sets that are not recursively enumerable. However, this only applies if the interpreting theory is itself recursively enumerable. The proof is implicit in (Orey, 1961) and (Feferman, 1960).

Lemma 4 (Formalized Compactness) *Let T be a consistent extension of $\mathrm{PA} + \mathrm{Tr}(\Pi_1^0)$ in $L[\mathrm{PA}]$ and Γ an arithmetically definable set of sentences of a relational language L of first-order logic with equality s.t. its deductive closure can be numerated in T. Let $\Gamma|k$ denote the set $\{\varphi \in \Gamma \mid \ulcorner\varphi\urcorner \leq k\}$, then $\forall k \in \mathbb{N} : \Gamma|k \prec T \Rightarrow \Gamma \prec T$.*

Proof. We assume that $\Gamma|k \prec T$ for every $k \in \mathbb{N}$ and that γ is a numeration of the deductive closure of Γ in T. As T is assumed to be consistent, $\Gamma|k$ is consistent for every $k \in \mathbb{N}$. Therefore, $T \vdash \mathrm{Con}_{\gamma_k}$ for every k (with γ_k in Σ_1^0 numerating the deductive closure of $\Gamma|k$ in T). Now define $\gamma^*(x) := \gamma(x) \wedge \mathrm{Con}_{\gamma_x}$ (x is free in Con_{γ_x}). Along the lines of the proof of (Lindström, 1997, Theorem 2.7) it follows that $T \vdash \mathrm{Con}_{\gamma^*}$. As γ numerates the deductive closure of Γ in T and $T \vdash \mathrm{Con}_{\gamma_k}$ for every $k \in \mathbb{N}$, also γ^* numerates the deductive closure of Γ in T. Finally, by $T \vdash \mathrm{Con}_{\gamma^*}$ and Lemma 2, we conclude that $\Gamma \prec T$. (Note that the Feferman-consistency statement Con_{γ^*} is equally eligible for an application of Lemma 2.) □

Proposition 2 *Let φ be a sentence of a relational language L of first-order logic with equality and Γ a set of L-sentences s.t. the deductive closure of Γ is arithmetically definable by an $L[\mathrm{PA}]$-formula in Π_n^0 (if $n > 0$, Σ_1^0 otherwise), then;*

1. $\exists M(structure_L(M) \wedge M \models \Gamma) \Leftrightarrow Sat^n(\Gamma)$

2. $\Gamma \models \varphi \Leftrightarrow LConseq^n(\Gamma, \varphi)$

3. $\models \varphi \Leftrightarrow LTr^*(\varphi)$

Proof. (1. $\Rightarrow$): Follows directly from Corollary 2. (1 $\Leftarrow$): Assume that $\Gamma \prec \mathrm{PA} + \mathrm{Tr}(\Pi^0_{n+1})$. Since $\mathrm{PA} + \mathrm{Tr}(\Pi^0_{n+1})$ is assumed to be consistent, Γ is also consistent and therefore satisfiable in some L-structure.

(2. $\Rightarrow$): Assume that $\Gamma \models \varphi$. Therefore, there exists a finite subset Γ' of Γ s.t. $\models \bigwedge \Gamma' \to \varphi$ and so (Lemma 1) $\forall f(subst(f) \Rightarrow \mathrm{PA} \vdash f(\bigwedge \Gamma' \to \varphi))$. Let f be any substitution function and assume that $\Gamma \prec_f \mathrm{PA} + \mathrm{Tr}(\Pi^0_{n+1})$. Therefore, $\mathrm{PA} + \mathrm{Tr}(\Pi^0_{n+1}) \vdash f(\bigwedge \Gamma')$ and $\mathrm{PA} + \mathrm{Tr}(\Pi^0_{n+1}) \vdash f(\varphi)$. So $LConseq^n(\Gamma, \varphi)$.

(2. $\Leftarrow$): Assume that $\Gamma \prec_f \mathrm{PA} + \mathrm{Tr}(\Pi^0_{n+1}) \Rightarrow \varphi \prec_f \mathrm{PA} + \mathrm{Tr}(\Pi^0_{n+1})$ for any substitution function f but $\Gamma \not\models \varphi$. $\Gamma \cup \{\neg\varphi\}$ is therefore consistent. The deductive closure of Γ is assumed to be arithmetically definable by a formula γ in Π^0_n. By Lemma 3, the deductive closure of Γ is bi-numerated by γ in $\mathrm{PA} + \mathrm{Tr}(\Pi^0_{n+1})$. Therefore, $\gamma(\ulcorner \neg \varphi \to \dot{x} \urcorner)$ (x is free in $\gamma(\ulcorner \neg \varphi \to \dot{x} \urcorner)$) numerates the deductive closure of $\Gamma \cup \{\neg\varphi\}$ in $\mathrm{PA} + \mathrm{Tr}(\Pi^0_{n+1})$. By Corollary 2, there exists a substitution function f s.t. $\Gamma \cup \{\neg\varphi\} \prec_f \mathrm{PA} + \mathrm{Tr}(\Pi^0_{n+1})$. In particular, $\Gamma \prec_f \mathrm{PA} + \mathrm{Tr}(\Pi^0_{n+1})$ so that it follows by assumption that $\mathrm{PA} + \mathrm{Tr}(\Pi^0_{n+1}) \vdash f(\varphi)$—but also $\mathrm{PA} + \mathrm{Tr}(\Pi^0_{n+1}) \vdash \neg f(\varphi)$, which is a contradiction as $\mathrm{PA} + \mathrm{Tr}(\Pi^0_{n+1})$ is assumed to be consistent. So $\Gamma \models \varphi$.

(3.): Follows from (2.) for $\Gamma = \emptyset$. □

Corollary 3 *Let φ be any sentence of a relational language L of first-order logic with equality and Γ a set of L-sentences s.t. the deductive closure of Γ is arithmetically definable by an $L[\mathrm{PA}]$-formula in Π^0_n (if $n > 0$, Σ^0_1 otherwise), then $\neg LConseq^n(\Gamma, \varphi) \Leftrightarrow Sat^n(\Gamma + \neg\varphi)$.*

The Corollary follows directly from Proposition 2 and it demonstrates that giving a model which satisfies $\Gamma + \neg\varphi$ in order to show that φ is not a logical consequence of Γ is equivalent to giving a substitution function f such that $\mathrm{PA} + \mathrm{Tr}(\Pi^0_{n+1}) \vdash f(\Gamma)$ but $\mathrm{PA} + \mathrm{Tr}(\Pi^0_{n+1}) \vdash \neg f(\varphi)$. In that respect, our substitutional definition of logical consequence is able to emulate an important property of its model-theoretic counterpart.

5 Conclusion

We proposed a substitutional definition of logical truth and consequence in terms of relative interpretations (Definition 2) and demonstrated that it is extensionally equivalent to the model-theoretic definitions for any relational first-order language with equality. We also demonstrated that the same result

Interpretations and Logical Consequence

can be obtained for a substitutional definition (Definition 3) that comes with a corresponding notion of satisfaction and compactness is not built-in to the notion of satisfaction—if we restrict the definition to arithmetically definable sets of assumptions. As the motivation of this undertaking was to eliminate the need for set-theory in the meta-theory of logical consequence, we will now have to review to what extent this has been successful.

First of all, the use of set-theory usually enters (also in our presentation) the meta-theory of logical consequence not first through the explanation of semantical notions but already in specifying syntactical terms like "language", "sentence" and "proof". However, it is well-known since Gödel that all of these syntactical notions can be adequately represented in a meta-theory not extending weak arithmetical theories.[18] The same holds for the notions of "translation function" and "relative interpretation". Apart from that, we obviously used set-theoretic terms like "set of sentences" in defining logical consequence substitutionally. If we want to specify this term purely arithmetically, we necessarily have to restrict the scope of our definition to sets of a certain complexity, namely to those which are definable by a first-order arithmetical formula. However, the notion of arithmetical definability itself involves the notion of satisfaction in the standard model of arithmetic. To avoid the use of set-theory in that aspect, we can instead restrict the definition to sets which can be numerated in an arithmetical theory. As Lemma 3 shows, this could be done in principle for any arithmetically definable set, but it requires a correspondingly strong arithmetical theory.

In the case of Definition 2, we don't need to accept, at first sight, a meta-theory that exceeds PA.[19] Leaving aside its discussed intensional flaws, if we try to not exceed PA in the framework of Definition 2, we have to admit that the class of sets which can be numerated in PA covers only the recursively enumerable ones. Without a doubt, this is a limitation compared to a model-theoretic setting. However, it could be argued that we don't lose much of generality since a non-recursively enumerable set of assumptions and its set of logical consequences is quite a rare thing to consider.

If we are interested in a substitutional definition that also meets the discussed intensional constraints, we necessarily have to exceed PA in any case. Even if we sacrifice the constraint that compactness should not be built-

---

[18] For details, see, e.g., (Smorynski, 1977).

[19] We assume that accepting a theory as a meta-theory implies that also its consistency is assumed and in case of an arithmetical theory also its ω-consistency, which is a common practice in Metamathematics. As pointed out, the 1-consistency of PA has to be assumed for Proposition 1 and the consistency of $PA + \text{Tr}(\Pi^0_{n+1})$ for Proposition 2.

in to the definition of satisfaction, we still need a meta-theory that involves $PA + Tr(\Pi_1^0)$ to allow for an adequate substitutional notion of satisfaction. It could be argued that this fact undermines our intend to eliminate the need for set-theory in the meta-theory of logical consequence as set-theory axiomatized by ZF is interpretable in $PA + Tr(\Pi_1^0)$. Still, we want to reply that Definition 3 offers a philosophical improvement for the meta-theory of logical consequence.

Evidently, we were able to replace the use of set-theoretic *vocabulary* by an arithmetical one. For the latter, it could be argued that its intuitive understanding is much clearer compared to the former as set-theory is a rather modern invention of 20th Century Mathematical Logic. Consequently, any definition in arithmetical terms is to favor over an equivalent definition in terms of sets.

Another point can be made if we consider again Quine's interest in a substitutional definition. He aimed to relieve the meta-theory of logical consequence from its ontological commitment to sets. The question is to what extent the fact that $ZF \prec PA + Tr(\Pi_1^0)$ affects this elimination of ontological commitment to sets. It has to be admitted that the term "ontological commitment" is not an entirely clear notion. So the previous question will not have a clear-cut answer and our argumentation can only be based on a somehow intuitive persuasion. Having said this, the aspect that we want to bring to attention here is that it is doubtful whether a presupposed ontology of a theory—in whatever sense—is preserved by relative interpretability. For instance, $ZF + CH \prec PA + Tr(\Pi_1^0)$ as well as $ZF + \neg CH \prec PA + Tr(\Pi_1^0)$. Intuitively, it seems plausible that an assumed ontology for $ZF + CH$ is in conflict with an assumed ontology for $ZF + \neg CH$. Thus, both ontologies cannot be adopted equally in an assumed ontology for $PA + Tr(\Pi_1^0)$. But as there appears to be no criterion to decide which one may be adopted, it seems questionable whether any ontology is preserved under relative interpretation in general.

Taking all this into consideration, we may conclusively say that this paper established a substitutional approach to logical truth and consequence that is able to avoid set-theoretic *vocabulary* in its meta-theory. For a definition that is extensionally adequate for recursively enumerable sets of assumptions, it was shown that the meta-theory has not to exceed PA. Though we don't expect our framework to replace the use of model-theory in the practice of determining a logical consequence, we consider it instructive that it could be done without set-theory in principle.

References

Boolos, G. S. (1975). On second-order logic. *Journal of Philosophy*, 72(16), 509–527.

Ebbs, G., & Goldfarb, W. (2018). First-order validity and the Hilbert-Bernays Theorem. *Philosophical Issues*, 28(1), 159–175.

Eder, G. (2016). Boolos and the metamathematics of Quine's definitions of logical truth and consequence. *History and Philosophy of Logic*, 32, 170–193.

Feferman, S. (1960). Arithmetization of metamathematics in a general setting. *Fundamenta Mathematicae*, 49(1), 35–92.

Halbach, V. (2018). The substitutional analysis of logical validity. *Nous*. doi: 10.1111/nous.12256.

Kaye, R. (1991). *Models of Peano Arithmetic*. Clarendon Press, Oxford.

Lindström, P. (1997). *Aspects of Incompleteness* (2nd (2003) ed.). Springer, Berlin.

Orey, S. (1961). Relative interpretations. *Mathematical Logic Quarterly*, 7(10), 146–153.

Quine, W. V. O. (1970). *Philosophy of Logic*. Harvard University Press, Cambridge, Massachusetts.

Smorynski, C. (1977). The incompleteness theorems. In J. Barwise (Ed.), *Handbook of Mathematical Logic* (pp. 821–866). North-Holland, Amsterdam.

Tarski, A. (1936). Über den Begriff der logischen Folgerung. *Actes du congres international de philosophie scientifique*, 7, 1–11.

Tarski, A. (1956). *Logic, Semantics, Metamathematics*. Clarendon Press, Oxford.

Tarski, A., Mostowski, A., & Robinson, R. M. (1953). *Undecidable theories*. North-Holland, Amsterdam.

Tarski, A., & Vaught, R. (1956). Arithmetical extensions of relational systems. *Compositio Mathematica*, 13, 81–102.

Wagner, H. (2019). Quine's substitutional definition of logical truth and the philosophical significance of the Löwenheim-Hilbert-Bernays theorem. *History and Philosophy of Logic*, 40(2), 182–199.

Mirko Engler
Humboldt University Berlin, Department of Philosophy
Germany
E-mail: mir.engler@gmail.com

The Problem of Harmony in Classical Logic

GIULIO GUERRIERI[1] AND ALBERTO NAIBO[2]

Abstract: A widely debated issue in philosophy of logic concerns the possibility of an inferentialist account of classical logic. Many proposals to show that classical logic satisfies the requirements of inferentialist semantics (such as harmony) demand to modify the ordinary natural deduction rules. In this paper, we try to explain why the ordinary natural deduction rules for classical logic are not harmonious and therefore not directly justifiable within an inferentialist framework. We show however that an indirect justification of classical logic, passing through negative translation, can be acceptable from an inferentialist point of view.

Keywords: harmony, inferentialism, classical logic, natural deduction, sequent calculus, negative translation, proof theory, philosophy of logic

1 Introduction

Inferentialism is an approach to the theory of meaning that identifies the meaning of an expression with its inferential relations with other expressions; it is in contradistinction to denotationalism, according to which denotations are the primary sort of meaning. The term *inferentialism* was coined by Brandom (2000), influenced by Wittgenstein and Sellars, as a label for his theory of language; it is also naturally applicable (and is growing increasingly common) within the philosophy of logic. In particular, this approach in semantics of logic aims to locate the meaning of propositions and logical connectives not in their interpretations, as in Tarskian approaches to semantics, but in the role that they play within a deductive system.

In the last few years, a widely debated issue in philosophy of logic is how to justify classical logic from an inferentialist point of view. Most of

---

[1] The work of the first author is partially funded by the EPSRC grant EP/R029121/1 "Typed Lambda-Calculi with Sharing and Unsharing".

[2] The work of the second author is partially funded by the ANR-DFG project *FFIUM – Formalism, Formalization, Intuition and Understanding in Mathematics* (ANR-17-FRAL-0003).

the works in such a line of research (e.g., Tennant, 1997) attempt to specify a set of inference rules for classical logic that fulfill the basic requirements of inferentialist semantics, such as the so called principle of *harmony* between introduction and elimination rules for logical constants. These works aim to show that inference rules of *ordinary* natural deduction for classical logic do not fulfill these requirements, but some *variants* of them do. For instance, this is the case for multi-conclusion inference rules in natural deduction (Read, 2000) and for a variant of natural deduction with two kinds of rules, corresponding to two different kinds of speech acts, assertion and denial (Rumfitt, 2000). But, quoting from Murzi (2020),[3]

> On the plausible assumption that our logical practice is both single-conclusion and assertion-based, it seemingly follows that classical logic, unlike intuitionistic logic, can't be accounted for in inferentialist terms.

So, classical logic as presented in ordinary natural deduction cannot be justified from an inferentialist viewpoint. This idea is supported by Prawitz (1977), Dummett (1991), Humberstone and Makinson (2012) among others.

In this paper, we aim to better understand the source of the issue: Why is ordinary (single-conclusion, assertion-based) classical natural deduction in trouble following an inferentialist approach? Why can ordinary classical natural deduction not be considered harmonious? To avoid a biased point of view, we adopt a concept of inferentialism as neutral as possible. In particular, we do not give priority to a kind of inference rules (e.g., introduction rules) over other kinds of inference rules (e.g., elimination rules).[4]

Moreover, we appeal to a translation of classical natural deduction into classical sequent calculus. In this way, we can make explicit the composition of derivations to understand better the structure of classical derivations.

Finally, we show that even though (unlike the intuitionistic case) reduction steps defined to prove the normalization theorem for ordinary classical natural deduction do not allow inference rules for classical negation to be seen as harmonious, however they allow us to settle an *indirect justification of classical logic*, via double negation translation.

---

[3] Murzi's (2020) inferentialist account will be discussed in Section 5.

[4] More precisely, in this work we want to focus on the inference rules used to construct a proof without giving priority to one kind of proofs over the other (e.g., proofs using an introduction rule as the last rule, instead of an elimination rule). In other words, we consider that an inferentialist semantics is not necessarily a proof-theoretic semantics, in which one has to specify what counts as a canonical proof (see Section 3).

2 Inferentialism and the harmony condition

In the philosophy of language and philosophy of logic, the starting point of *inferentialism* (or inferentialist semantics) is that the meaning of our linguistic expressions and, more generally, of our concepts is given by the way in which we use them in our linguistic practice. This position (for an overview, see Peregrin, 2012, 2014) is inspired by the later work of Wittgenstein about "meaning as use", in particular from his *Philosophical Investigations*:

> For a large class of cases of the employment of the word 'meaning'—though not for all—this word can be explained in this way: the meaning of a word is its use in the language. (§ 43)

This idea is then specified by the account of inference given by Sellars (e.g., Sellars, 1953): the meaning of an expression is determined by the way we use that expression in our inferential practice. The term inferentialism was eventually coined by Brandom (2000, p. 11) as a label for his theory of language, conceived as the game of "giving and asking for reasons":

> Grasping the concept ... is mastering its inferential use: knowing (in the practical sense of being able to distinguish, a kind of knowing how) what else one would be committing oneself to by applying the concept, what would entitle one to do so, and what would preclude such entitlement.

As a theory of meaning, inferentialism is opposed to semantic denotationalism (or referentialism), according to which denotations are the primary sort of meaning. Still quoting from Brandom (2000):

> [Consider] a general model of conceptual contents as inferential roles that has been recommended by Dummett. According to that model, the use of any linguistic expression or concept has two aspects: the *circumstances* under which it is correctly applied, uttered, or used, an the appropriate *consequences* of its application, utterance, or use. (p. 62)

In philosophy of logic, the inferentialist standpoint claims that the meaning of logical constants is a matter of their *inference rules*, which govern proofs. According to Tennant (2007),

> An inferentialist theory of meaning holds that the meaning of a logical operator can be captured by suitably formulated rules of inference (in, say, a system of natural deduction). (p. 1056)

Giulio Guerrieri and Alberto Naibo

We can thus reformulate the distinction between circumstances and consequences mentioned above in the framework of *natural deduction*: the *introduction rules* for a logical constant correspond to the circumstances under which it is correctly used, and the *elimination rules* of a logical constant correspond to the appropriate consequences that can be inferred from its use. For instance, in a deduction system such as natural deduction, the meaning of the conjunction ($\wedge$) is fixed by its introduction and elimination rules:

$$\frac{A \quad B}{A \wedge B} \wedge_{\text{intro}} \qquad \frac{A \wedge B}{A} \wedge_{\text{elim}_1} \qquad \frac{A \wedge B}{B} \wedge_{\text{elim}_2}$$

In this way the semantics of logical constants is given in terms of proofs (in a deduction system) instead of truth-conditions.

However, it is not enough to specify the inference rules for a logical constant to determine its meaning. Indeed, as shown by Prior (1960), if we introduce a logical constant *tonk* whose inference rules are

$$\frac{A}{A \text{ tonk } B} \text{ tonk}_{\text{intro}} \qquad \frac{A \text{ tonk } B}{B} \text{ tonk}_{\text{elim}}$$

then we can infer any statement B from any statement A:

$$\frac{\dfrac{A}{A \text{ tonk } B} \text{ tonk}_{\text{intro}}}{B} \text{ tonk}_{\text{elim}}$$

In particular, if the inference rules for tonk are added to intuitionistic or classical logic, we can derive a contradiction ($\bot$):

$$\frac{\dfrac{\dfrac{[A]^1}{A \to A} \to_{\text{intro}} 1}{(A \to A) \text{ tonk } \bot} \text{ tonk}_{\text{intro}}}{\bot} \text{ tonk}_{\text{elim}}$$

So, apparently harmless inference rules may make inconsistent our inferential practice, and make meaningless the logical constant they define.

In order to rely the meaning of a logical constant on its inference rules, an inferentialist should require something more (see Belnap, 1962; Dummett, 1991): a *harmony* between the circumstances we can use a logical constant and the consequences that we can infer from its use. Technically, in proof-theory the *principle of harmony* states that an introduction rule and an elimination rule should "cancel out" in the sense that if in a derivation a logical constant is introduced and then immediately eliminated (a so called *detour*), there is no gain. For example, we can define the following rewrite steps (noted with $\rightsquigarrow$) to cancel out detours for the conjunction:

The Problem of Harmony in Classical Logic

$$\dfrac{\dfrac{A \quad B}{A \wedge B}\wedge_{\text{intro}}}{A}\wedge_{\text{elim}_1} \rightsquigarrow A \qquad \dfrac{\dfrac{A \quad B}{A \wedge B}\wedge_{\text{intro}}}{B}\wedge_{\text{elim}_2} \rightsquigarrow B$$

Inferentialism then requires that for all logical constants the harmony between their introduction and elimination rules in natural deduction should be satisfied. In particular, in intuitionistic first-order logic, showing—as done by Prawitz (1965, Ch. IV)—that the rules for all usual connectives and quantifiers are harmonious allows one to prove the normalization theorem for this logic (i.e., all the detours in a derivation can be avoided), and also to obtain its consistency as a corollary, which guarantees that the logical constants defined by their introduction and elimination rules are meaningful.

3   Harmony and classical logic

In (propositional and first-order) natural deduction, classical logic can be obtained by adding the elimination rule for the double negation ($\neg\neg_{\text{elim}}$)

$$\dfrac{\neg\neg A}{A}\,\neg\neg_{\text{elim}}$$

to the set of inference rules for intuitionistic logic. However, such a deduction system breaks harmony because a detour like

$$\dfrac{\dfrac{[\neg A]^1}{\vdots}}{\dfrac{\bot}{\neg\neg A}\,\neg_{\text{intro}}\,1}\,\neg\neg_{\text{elim}}$$
$$\dfrac{}{A}$$

(which also breaks the subformula property) cannot be canceled out.

Another way to extend (propositional and first-order) natural deduction from intuitionistic logic to classical logic is to add the inference rule for *reductio ad absurdum* (raa) instead of $\neg\neg_{\text{elim}}$

$$\dfrac{\dfrac{[\neg A]^1}{\vdots}}{\dfrac{\bot}{A}\,\text{raa}\,1} \tag{1}$$

In this way, it is possible to define rewrite rules to cancel out a new kind of detour introduced by raa, and to prove normalization (and then consistency) of classical natural deduction. Following Prawitz (1965), when the conclusion A of raa is a non-atomic formula, raa can be seen as an introduction rule for the principal connective of A. So, we have a new kind of detours, the so called *classical detours*, when in a derivation a non-atomic formula occurrence A is both the conclusion of raa and the premise of an elimination rule for the principal connective of A. The following rewrite step shows how to cancel out the classical detour for the conjunction $\wedge$:

$$\begin{array}{c}
[\neg(A \wedge B)]^1 \\
\vdots \\
\dfrac{\bot}{A \wedge B} \text{ raa 1} \\
\dfrac{}{A} \wedge_{\text{elim}} \\
\vdots \\
C
\end{array}
\quad \leadsto \quad
\begin{array}{c}
\dfrac{\dfrac{[A \wedge B]^1}{A} \wedge_{\text{elim}} \quad [\neg A]^2}{\dfrac{\bot}{\neg(A \wedge B)} \neg_{\text{intro 1}}} \neg_{\text{elim}} \\
\vdots \\
\dfrac{\bot}{A} \text{ raa 2} \\
\vdots \\
C
\end{array}
\qquad (2)$$

Analogous rewrite rules can be defined for all the usual logical constants in first-order classical natural deduction, so as to prove normalization (see Stålmarck, 1991; von Plato & Siders, 2012). Is it enough to conclude that classical natural deduction is harmonic? Usually in the literature the answer is negative. Such a negative answer relies on two assumptions:

1. some inference rules (for instance, the introduction ones) have priority over other inference rules (for instance, the elimination ones), in the sense that the former are primitive and self-justified;

2. the meaning of a sentence is determined not only by the inference rules but also by the fact that there is a canonical proof of such a statement, that is a proof whose last rule is primitive and self-justified (for instance, an introduction rule).

This is Dummett's standpoint (see Dummett, 1991, among others), which is actually a commitment to something more than mere inferentialism, because it requires also a semantics of proofs, such as the one developed in proof-theoretic semantics (Schroeder-Heister, 2018).

We aim to show that the inference rules for ordinary classical (propositional and first-order) natural deduction cannot be justified not only in proof-theoretic semantics but also from the mere inferentialist point of view, which simply requires (see Murzi, 2020) that our inferential practice is based

The Problem of Harmony in Classical Logic

on one type only of speech act, i.e., single-conclusion assertions. In order to do this, we show in the next section that classical detours are not real detours, in the sense that these configurations cannot be seen as an intro/elim pattern. This impossibility is more evident when we analyze classical detours in *sequent calculus* instead of natural deduction. Technically, there are essentially two reasons to move to sequent calculus.

1. Sequent calculus allows us to decompose the structure of a derivation in natural deduction into sub-derivations whose composition is made explicit by the cut rule. In this way, modularity and compositionality characterizing the derivation construction process are made evident.

2. As said in Section 1, the proposals that try to provide an inferentialist account of classical logic introduce some features of sequent calculus in natural deduction, e.g., inference rules with several conclusions (like in a sequent calculus with many formulas on the right-hand side of a sequent), or two kinds of speech acts, assertions and denials (like in a sequent calculus where inference rules modify formulas both on the left-hand side and on the right-hand side of a sequent, so that inference rules act on formulas with two distinct "polarities"). See also Restall (2005, 2008).

Moving from natural deduction to sequent calculus is then, in our opinion, a more indulgent and less committed way to consider classical logic. Therefore, if we manage to show that a classical detour is not a real detour, not even in the sequent calculus framework, our argument against a *direct* inferentialist justification of classical logic is stronger than the ones based on classical natural deduction.

4 Harmony and cut

Our starting point is the observation that by adopting a suitable translation from *intuitionistic* natural deduction to sequent calculus (like the one presented by von Plato, 2003, 2011), it is possible to show that, given a certain (intuitionistic) connective c, a c-detour corresponds to a cut where the cut formula in both premises is principal (i.e., it comes from a rule for the connective c). In the case of conjunction, the $\wedge$-detour in natural deduction

$$
\begin{array}{c}
\Gamma \quad \Delta \\
\vdots \quad \vdots \\
\dfrac{\dfrac{A \quad B}{A \wedge B}\wedge_{\text{intro}}}{A}\wedge_{\text{elim}_1} \quad \text{corresponds to} \quad \dfrac{\dfrac{\dfrac{\Gamma \vdash A \quad \Delta \vdash B}{\Gamma, \Delta \vdash A \wedge B}\wedge_R \quad \dfrac{\Gamma, \Delta, A \vdash C}{\Gamma, \Delta, A \wedge B \vdash C}\wedge_{L_1}}{\dfrac{\Gamma, \Gamma, \Delta, \Delta \vdash C}{\Gamma, \Delta \vdash C}\text{Ctr}_L}}{\text{Cut}} \\
\vdots \\
C
\end{array}
$$

in the sequent calculus, where $\wedge_R$ and $\wedge_L$ are the right and left rule for conjunction, respectively; while Ctr_L indicates the left-contraction rule.

Therefore, in first-order intuitionistic logic, a detour in natural deduction corresponds to a so called *key-case* in the cut elimination procedure for sequent calculus, and the *harmony* condition becomes the possibility to *reduce the key-cases*. Harmony is thus just a special case of cut elimination.

Our claim is that to determine whether *classical* logic satisfies or not the harmony condition, a good reasonable test is to check whether a classical detour (as defined in Section 3) creates a key-case cut (for which we possess an elimination strategy), once it is translated into sequent calculus.

To operate such a test, we have first to translate the rule raa in (1) into the classical sequent calculus. What we obtain is the following configuration:

$$
\dfrac{\dfrac{\Gamma, \neg A \vdash}{\Gamma \vdash \neg\neg A}\neg_R \quad \dfrac{\dfrac{\dfrac{A \vdash A}{\vdash A, \neg A}\neg_R}{\neg\neg A \vdash A}\neg_L}{\Gamma \vdash A}\text{Cut}
$$

For example, when it is translated into the sequent calculus, the classical detour considered on the left-hand side of (2) becomes:

$$
\dfrac{\dfrac{\dfrac{\Gamma, \neg(A \wedge B) \vdash}{\Gamma \vdash \neg\neg(A \wedge B)}\neg_R \quad \dfrac{\dfrac{A \wedge B \vdash A \wedge B}{\vdash A \wedge B, \neg(A \wedge B)}\neg_R}{\neg\neg(A \wedge B) \vdash A \wedge B}\neg_L}{\Gamma \vdash A \wedge B}\text{Cut}_1 \quad \dfrac{\Gamma, A \vdash C}{\Gamma, A \wedge B \vdash C}\wedge_{L_1}}{\dfrac{\Gamma, \Gamma \vdash C}{\Gamma \vdash C}\text{Ctr}_L}\text{Cut}_2 \quad (3)
$$

Note in fact that our translation (3) creates a key-case cut with cut formula $\neg\neg(A \wedge B)$ (Cut_1). However, the formula that "disappears" after the rewrite step of the classical detour considered in (2) is $A \wedge B$, which corresponds

The Problem of Harmony in Classical Logic

to the cut formula for Cut_2 in our translation (3) into the sequent calculus. But Cut_2 is *not* a key-case. Can some operations on (3) transform Cut_2 into a key-case? We can follow Urban (2001) and consider the possibility to let Cut_2 pass over the Cut_1,[5] so as to obtain:

$$\cfrac{\cfrac{\vdots}{\cfrac{\Gamma, \neg(A \wedge B) \vdash}{\Gamma \vdash \neg\neg(A \wedge B)}\neg_R} \quad \cfrac{\cfrac{\cfrac{\overline{A \wedge B \vdash A \wedge B}^{ax}}{\vdash A \wedge B, \neg(A \wedge B)}\neg_R}{\neg\neg(A \wedge B) \vdash A \wedge B}\neg_L \quad \cfrac{\vdots \quad \Gamma, A \vdash C}{\Gamma, A \wedge B \vdash C}\wedge_{L_1}}{\Gamma, \neg\neg(A \wedge B) \vdash C}Cut_2}{\cfrac{\Gamma, \Gamma \vdash C}{\Gamma \vdash C}Ctr_L}Cut_1 \quad (4)$$

Still, Cut_2 is not a key-case for the cut formula $A \wedge B$, according to the cut elimination procedure. However, it could be argued that such a key-case can be obtained by operating an *expansion* on the axiom sequent $A \wedge B \vdash A \wedge B$ in (4), that is, by replacing it with the derivation:

$$\cfrac{\cfrac{\overline{A \vdash A}^{ax}}{A \wedge B \vdash A}\wedge_{L_1} \quad \cfrac{\overline{B \vdash B}^{ax}}{A \wedge B \vdash B}\wedge_{L_2}}{A \wedge B \vdash A \wedge B}\wedge_R$$

In this way, by permuting Cut_2 upward, one finally reaches a key-case in which the premise of the cut comes from the $\wedge_R$ of the expanded derivation and the $\wedge_{L_1}$ used to derive $\Gamma, A \wedge B \vdash C$, i.e.,

$$\cfrac{\cfrac{\vdots}{\cfrac{\Gamma, \neg(A \wedge B) \vdash}{\Gamma \vdash \neg\neg(A \wedge B)}\neg_R} \quad \cfrac{\cfrac{\cfrac{\cfrac{\overline{A \vdash A}^{ax}}{A \wedge B \vdash A}\wedge_{L_1} \quad \cfrac{\overline{B \vdash B}^{ax}}{A \wedge B \vdash B}\wedge_{L_2}}{A \wedge B \vdash A \wedge B}\wedge_R \quad \cfrac{\vdots \quad \Gamma, A \vdash C}{\Gamma, A \wedge B \vdash C}\wedge_{L_1}}{\cfrac{\Gamma, A \wedge B \vdash C}{\Gamma \vdash C, \neg(A \wedge B)}\neg_R}Cut_2}{\Gamma, \neg\neg(A \wedge B) \vdash C}\neg_L}{\cfrac{\Gamma, \Gamma \vdash C}{\Gamma \vdash C}Ctr_L}Cut_1 \quad (5)$$

---

[5]As Urban (2001, p. 417) remarks, this operation of switching the order of the cuts in the cut elimination procedure allows one to preserve "in the intuitionistic case the correspondence between cut-elimination and beta-reduction". Now, under the so-called Curry–Howard correspondence, one step of beta-reduction (in typed lambda-calculus) corresponds to one step of detour elimination (in natural deduction). And since the harmony condition corresponds indeed to the possibility of executing one step of detour elimination, it is then reasonable to admit this cut-switching operation, as it allow us to be more faithful with respect to the inferentialist requirements in our translation from natural deduction into sequent calculus.

Moreover, Cut$_2$ can be eliminated, so as to obtain the derivation:

$$
\cfrac{
  \cfrac{
    \cfrac{\Gamma, \neg(A \wedge B) \vdash}{\Gamma \vdash \neg\neg(A \wedge B)} \neg R
    \qquad
    \cfrac{
      \cfrac{
        \cfrac{\vdots \quad \Gamma, A \vdash C}{\Gamma, A \wedge B \vdash C} \wedge L_1
      }{\Gamma \vdash C, \neg(A \wedge B)} \neg R
    }{\Gamma, \neg\neg(A \wedge B) \vdash C} \neg L
  }{\Gamma, \Gamma \vdash C} \text{Cut}_1
}{\Gamma \vdash C} \text{Ctr}_L
$$

Can this argument be used in order claim that classical logic is indeed harmonious? At first sight, it seems that some result has been achieved, since a key-case of a cut has been eliminated. However, this is just a false impression, because the key-case that has been eliminated (in the sequent calculus) does not correspond to the the classical detour (in natural deduction) considered on the left-hand side of (2). This is due to the fact that in the sequent calculus one can create cuts that do not necessarily correspond to detours in natural deduction.

To understand this point, let us proceed backwards, and translate into natural deduction the sub-derivation of (5) with conclusion $\Gamma, A \wedge B \vdash C$ and last rule Cut$_2$: one gets the derivation (with a $\wedge$-detour)

$$
\cfrac{
  \cfrac{
    \cfrac{A \wedge B}{A} \wedge_{\text{elim}_1} \qquad \cfrac{A \wedge B}{B} \wedge_{\text{elim}_2}
  }{A \wedge B} \wedge_{\text{intro}}
}{A} \wedge_{\text{elim}_1}
$$
$$
\vdots
$$
$$
C
$$

and if we now plug the derivation below over the hypothesis $A \wedge B$

$$
\Gamma, [\neg(A \wedge B)]^1
$$
$$
\vdots
$$
$$
\cfrac{\bot}{A \wedge B} \text{ raa } 1
$$

we obtain the derivation (in natural deduction) corresponding to (5), and it is *not* the derivation on the right-hand side of (2), in particular it still contains the classical detour of the derivation on the left-hand side of (2):

The Problem of Harmony in Classical Logic

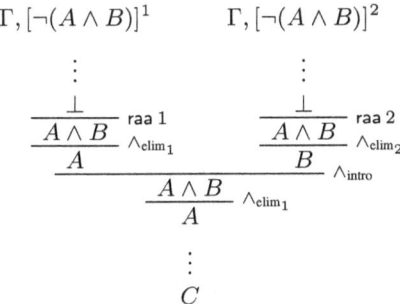

The ∧-detour present in this derivation can be eliminated, but its elimination does not involve the raa rule. Moreover, by eliminating it, we just get back to the original derivation on the left-hand side of (2). This means that the operation of expansion we considered is absolutely useless and does not play any relevant role with respect to the question of harmony.

5 Conclusion: classical reduction and negative translations

The analysis of the raa rule that we proposed in the previous section seems to confirm Dummett's (1991) point of view, according to which inferentialism entails a form of logical revisionism. In particular, classical logic should be rejected since the failure of harmony for the inference rules that distinguish it from intuitionistic logic (such as raa) prevents one from conferring a coherent meaning to logical constants.[6] However, our opinion is that it is still possible to use reduction procedures associated with classical rules—and in particular with raa—in order to assign a meaning to classical logic from the inferentialist point of view. More precisely, the idea is that a negative translation from classical to intuitionistic logic can be extracted from the reduction procedure for raa (as suggested for instance by Pereira, 2000). In this way, even if the explanation of the meaning of classical sentences is an *indirect* one (since it passes through a translation), yet it rests on inferential considerations (since it rests on a reduction procedure).

---

[6]This idea is often sustained by noting that classical logic is a non-conservative extension of intuitionistic logic. A coherent theory of meaning should explain the meaning of a sentence simply by making reference to the expressions that compose that sentence (what Dummett, 1975, p. 12, calls a "molecular view of language"). However, if we consider the purely implicative sentence $((A \to B) \to A) \to A$, it is not possible to give an account of its meaning only in terms of the inference rules for the implication (which are already part of the rules of intuitionistic logic), since this sentence is derivable only by using classical rules.

Let us start noting that the reduction procedure for the classical detour we mentioned above is in fact a particular case of a more general procedure defined by Prawitz (1965, p. 40) and consisting in "atomizing" the occurrences of the raa rule. For instance, if we consider a derivation like

$$\Gamma, [\neg(P \wedge Q)]^1$$
$$\vdots$$
$$\frac{\bot}{P \wedge Q} \text{ raa 1}$$

where P and Q are atomic formulas, then, according to Prawitz, it is possible to transform it into a derivation where raa is applied to obtain P and Q instead of the non-atomic formula $P \wedge Q$, i.e.,

$$\cfrac{\Gamma, \cfrac{[\neg P]^2 \quad \cfrac{\cfrac{[P \wedge Q]^1}{P} \wedge_{\text{elim}_1}}{\neg(P \wedge Q)} \neg_{\text{elim}}}{\vdots} \neg_{\text{intro } 1}}{\cfrac{\bot}{P} \text{ raa 2}} \quad \cfrac{\Gamma, \cfrac{[\neg Q]^4 \quad \cfrac{\cfrac{[P \wedge Q]^3}{Q} \wedge_{\text{elim}_1}}{\neg(P \wedge Q)} \neg_{\text{elim}}}{\vdots} \neg_{\text{intro } 3}}{\cfrac{\bot}{Q} \text{ raa 4}} \wedge_{\text{intro}}$$
$$P \wedge Q$$

Now, if these two occurrences of raa are the only occurrences of raa in the derivation, then we can transform it into an intuitionistic derivation simply by replacing each of these two occurrences of raa by a $\neg_{\text{intro}}$ rule:

$$\cfrac{\Gamma, \cfrac{[\neg P]^2 \quad \cfrac{\cfrac{[P \wedge Q]^1}{P} \wedge_{\text{elim}_1}}{\neg(P \wedge Q)} \neg_{\text{elim}}}{\vdots} \neg_{\text{intro } 1}}{\cfrac{\bot}{\neg\neg P} \neg_{\text{intro } 2}} \quad \cfrac{\Gamma, \cfrac{[\neg Q]^4 \quad \cfrac{\cfrac{[P \wedge Q]^3}{Q} \wedge_{\text{elim}_1}}{\neg(P \wedge Q)} \neg_{\text{elim}}}{\vdots} \neg_{\text{intro } 3}}{\cfrac{\bot}{\neg\neg Q} \neg_{\text{intro } 4}} \wedge_{\text{intro}}$$
$$\neg\neg P \wedge \neg\neg Q$$

The previous operation transforms atomic formulas into their double negation. This is exactly what is done by the *Gödel-Gentzen negative translation* (see Ferreira & Oliva, 2012). The problem is that Prawitz's atomization

The Problem of Harmony in Classical Logic

procedure for raa can be applied only to conjunction, implication, and the universal quantifier. If we want to extract the Gödel-Gentzen negative translation from the reduction procedure for raa, we cannot then take disjunction and the existential quantifier as primitive; instead, we have to define them from the very beginning in terms of negation plus conjunction and the universal quantifier, respectively. But in this way the Gödel-Gentzen negative translation for disjunction and the existential quantifier will not be justified by a proof-transformation procedure, but only by a stipulation. It is indeed for this reason that Murzi (2020) presents a system for classical natural deduction in which he replaces the standard rules for disjunction by adopting a new kind of rules which work modulo the definition of disjunction in terms of negation and conjunction (i.e., $A \vee B =_{def} \neg(\neg A \wedge \neg B))$[7].

However, if we restrict our analysis to the propositional case, we can find a justification of a negative translation from the reduction procedure of raa for the full propositional fragment. The idea is to work with the variant of Seldin's *reduction steps for* raa presented by Guerrieri and Naibo (2019).[8] This reduction strategy can be seen as the dual of Prawitz's one: while Prawitz's procedure breaks down classical reasoning into a number of atomic steps of raa, Seldin's procedure compactifies classical reasoning into one single (possibly complex) step of raa, the final one. What is remarkable is that Seldin's strategy can be applied also in the case of disjunction. Consider for instance the case in which the major premise of a $\vee_{elim}$ rule is the conclusion of a raa rule:

$$\cfrac{\cfrac{\Gamma, [\neg(A \vee B)]^1}{\vdots} \quad \Delta, [A]^2 \quad \Sigma, [B]^2}{\cfrac{\bot}{A \vee B} \text{ raa 1} \quad \cfrac{\vdots}{C} \quad \cfrac{\vdots}{C}} \vee_{elim} 2$$

According to Seldin's reduction, this derivation can be transformed into

---

[7]We prefer not to present here Murzi's rules for disjunction because, strictly speaking, they do not operate on formulas, but on rules. They are thus higher-level rules, and in order to present them we should first explain how a system of higher-level rules work. However, due to space constraints, we cannot do it here. A similar version of Murzi's rule for disjunction was already introduced by Prawitz (2015), and discussed by Pereira and Rodriguez (2017).

[8]These reduction steps (both the original presentation due to Seldin, 1986 and its variant introduced by Guerrieri & Naibo, 2019) push downwards the rule raa in a derivation in classical natural deduction, so as to transform it in a derivation with same premises and conclusion and at most one occurrence of raa, and this occurrence (if any) is its last rule.

$$
\begin{array}{c}
\Gamma, \quad \cfrac{[A \vee B]^2 \quad \cfrac{\cfrac{\Delta, [A]^1}{\vdots}}{\cfrac{[\neg C]^3 \quad C}{\bot}\,\neg_{\text{elim}}} \quad \cfrac{\cfrac{\Sigma, [B]^1}{\vdots}}{\cfrac{[\neg C]^3 \quad C}{\bot}\,\neg_{\text{elim}}}}{\cfrac{\bot}{\neg(A \vee B)}\,\neg_{\text{intro}}\,2}\,\vee_{\text{elim}}\,1 \\
\vdots \\
\cfrac{\bot}{C}\,\text{raa}\,3
\end{array}
$$

We can then replace the last occurrence of raa by a $\neg_{\text{intro}}$ rule and obtain:

$$
\begin{array}{c}
\Gamma, \quad \cfrac{[A \vee B]^2 \quad \cfrac{\cfrac{\Delta, [A]^1}{\vdots}}{\cfrac{[\neg C]^3 \quad C}{\bot}\,\neg_{\text{elim}}} \quad \cfrac{\cfrac{\Sigma, [B]^1}{\vdots}}{\cfrac{[\neg C]^3 \quad C}{\bot}\,\neg_{\text{elim}}}}{\cfrac{\bot}{\neg(A \vee B)}\,\neg_{\text{intro}}\,2}\,\vee_{\text{elim}}\,1 \\
\vdots \\
\cfrac{\bot}{\neg\neg C}\,\neg_{\text{intro}}\,3
\end{array}
$$

Moreover, Seldin's procedure can be applied also when the application of the raa rule is followed not by an elimination rule, but by an introduction one. For instance, the derivation

$$
\begin{array}{c}
\Gamma, [\neg A]^1 \\
\vdots \\
\cfrac{\cfrac{\bot}{A}\,\text{raa}\,1}{A \vee B}\,\vee_{\text{intro}_1}
\end{array}
\qquad \text{can be transformed into} \qquad
\begin{array}{c}
\Gamma, \quad \cfrac{[\neg(A \vee B)]^1 \quad \cfrac{[A]^2}{A \vee B}\,\vee_{\text{intro}_1}}{\cfrac{\bot}{\neg A}\,\neg_{\text{intro}}\,2}\,\neg_{\text{elim}} \\
\vdots \\
\cfrac{\bot}{A \vee B}\,\text{raa}\,1
\end{array}
$$

Again, by replacing the last raa rule on the right hand-side proof with a $\neg_{\text{intro}}$ rule, we obtain an intuitionistic proof of $\neg\neg(A \vee B)$. Now, when we restrict to the propositional case, translating a formula simply by putting a double negation in front of it corresponds to the so-called *Kuroda's negative translation* (see Ferreira & Oliva, 2012). Therefore, if we restrict to the propositional case, we have that Seldin's reduction procedure for raa can be used to justify an intuitionistic (and so acceptable from an inferentialist viewpoint) interpretation of classical sentences by means of Kuroda's translation (see Guerrieri & Naibo, 2019). What is remarkable is that the translation is obtained by a transformation induced by a reduction procedure for classical proofs. It can thus be argued that classical logic, as well as its reduction procedures, is not devoid of any inferential meaning: it possesses indeed an inferential meaning, but this meaning has to be extracted by adopting an intuitionistic point of view.

The limit of this argument is that it cannot be applied in an *uniform* way beyond the propositional case. On the one hand, as we have already said, Prawitz's classical reductions can be applied in the case of the universal quantifier, but they cannot be applied in the case the existential quantifier. Therefore, it is not possible to induce a translation for existential classical sentences from these reductions. On the other hand, Seldin's classical reductions can be applied in the case of the existential quantifier, but not in the case of the universal quantifier (see Guerrieri & Naibo, 2019; Seldin, 1986). Thus, no translation for universal classical sentences can be induced from Seldin's reductions. Does this lack of uniformity in the treatment of quantifiers represent a limit to the idea that the meaning of classical operators can be understood inferentially by means of inferentially justified negative translations? We leave this question open to further investigations.

References

Belnap, N. D. (1962). Tonk, plonk and plink. *Analysis*, 22(6), 130–134.
Brandom, R. (2000). *Articulating Reasons: An Introduction to Inferentialism*. Harvard University Press.
Dummett, M. (1975). The philosophical basis of intuitionistic logic. In H. E. Rose & J. C. Shepherdson (Eds.), *Logic Colloquium '73* (pp. 5–40). Amsterdam: North-Holland.
Dummett, M. (1991). *The Logical Basis of Metaphysics*. Harvard University Press.
Ferreira, G., & Oliva, P. (2012). On the relation between various negative translations. In U. Berger, H. Diener, P. Schuster, & M. Seisenberger (Eds.), *Logic, Construction, Computation* (pp. 227–258). De Gruyter.

Guerrieri, G., & Naibo, A. (2019). Postponement of raa and Glivenko's theorem, revisited. *Studia Logica, 107*(1), 109–144.

Humberstone, L., & Makinson, D. (2012, 02). Intuitionistic logic and elementary rules. *Mind, 120*(480), 1035–1051.

Murzi, J. (2020). Classical harmony and separability. *Erkenntnis, 85*(2), 391–415. doi: 10.1007/s10670-018-0032-6

Peregrin, J. (2012). What is inferentialism? In L. Gurova (Ed.), *Inference, Consequence and Meaning (Perspectives on Inferentialism)*, (pp. 3–16). Cambridge Scholars Publishing.

Peregrin, J. (2014). *Inferentialism: Why rules matter*. Palgrave Macmillan UK.

Pereira, L. C. (2000). Translations and normalization procedures (abstract). In R. J. G. B. de Queiroz (Ed.), *WoLLIC'2000 – 7th Workshop on Logic, Language, Information and Computation* (pp. 21–24). www.cin.ufpe.br/~wollic/wollic2000/proceedings/anais.ps.gz.

Pereira, L. C., & Rodriguez, R. O. (2017). Normalization, soundness and completeness for the propositional fragment of Prawitz's ecumenical system. *Revista Portuguesa de Filosofia, 73*(3–4), 1153–1168.

Prawitz, D. (1965). *Natural Deduction: A proof-theoretical study*. Almqvist & Wiksell.

Prawitz, D. (1977). Meaning and proofs: On the conflict between classical and intuitionistic logic. *Theoria, 43*(1), 2–40.

Prawitz, D. (2015). Classical versus intuitionistic logic. In E. H. Haeusler, W. de Campos Sanz, & B. Lopes (Eds.), *Why is this a Proof? Festschrift for Luiz Carlos Pereira* (pp. 15–32). College Publications.

Prior, A. N. (1960). The runabout inference-ticket. *Analysis, 21*(2), 38–39.

Read, S. (2000). Harmony and autonomy in classical logic. *Journal of Philosophical Logic, 29*(2), 123–154.

Restall, G. (2005). Multiple conclusions. In P. Hájek, L. Valdés-Villanueva, & D. Westerståhl (Eds.), *Logic, Methodology and Philosophy of Science: Proceedings of the Twelfth International Congress* (pp. 189–205). London: King's College Publications.

Restall, G. (2008). Assertion and denial, commitment and entitlement, and incompatibility (and some consequence). *Studies in Logic, 1*, 26–36.

Rumfitt, I. (2000). 'Yes and no'. *Mind, 109*(436), 781–823.

Schroeder-Heister, P. (2018). Proof-theoretic semantics. In E. N. Zalta (Ed.), *The Stanford Encyclopedia of Philosophy* (Spring 2018 Edition). https://plato.stanford.edu/archives/spr2018/entries/proof-theoretic-semantics/.

Seldin, J. P. (1986). On the proof theory of the intermediate logic MH. *The Journal of Symbolic Logic*, *51*(3), 626–647.

Sellars, W. (1953). Inference and meaning. *Mind*, *62*(247), 313–338.

Stålmarck, G. (1991). Normalization theorems for full first order classical natural deduction. *The Journal of Symbolic Logic*, *56*(1), 129–149.

Tennant, N. (1997). *The Taming of the True*. Oxford University Press.

Tennant, N. (2007). Existence and identity in free logic: A problem for inferentialism? *Mind*, *116*(464), 1055–1078.

Urban, C. (2001). Strong normalisation for a Gentzen-like cut-elimination procedure. In S. Abramsky (Ed.), *Typed Lambda Calculi and Applications* (pp. 415–429). Springer.

von Plato, J. (2003). Translations from natural deduction to sequent calculus. *Mathematical Logic Quarterly*, *49*(5), 435–443.

von Plato, J. (2011). A sequent calculus isomorphic to Gentzen's natural deduction. *The Review of Symbolic Logic*, *4*(1), 43–53.

von Plato, J., & Siders, A. (2012). Normal derivability in classical natural deduction. *The Review of Symbolic Logic*, *5*(2), 205–211.

Giulio Guerrieri
University of Bath, Department of Computer Science
United Kingdom
E-mail: g.guerrieri@bath.ac.uk

Alberto Naibo
Université Paris 1 Panthéon-Sorbonne, Department of Philosophy,
Institut d'Histoire et de Philosophie des Sciences et des Techniques
(IHPST, UMR 8590)
France
E-mail: alberto.naibo@univ-paris1.fr

Expressing Validity: Towards a Self-Sufficient Inferentialism

ULF HLOBIL[1]

Abstract: For semantic inferentialists, the basic semantic concept is validity. An inferentialist theory of meaning should offer an account of the meaning of "valid." If one tries to add a validity predicate to one's object language, however, one runs into problems like the v-Curry paradox. In previous work, I presented a validity predicate for a non-transitive logic that can adequately capture its own meta-inferences. Unfortunately, in that system, one cannot show of any inference that it is invalid. Here I extend the system so that it can capture invalidities.

Keywords: inferentialism, naive validity, v-Curry paradox, non-transitive consequence, substructural logic

1 Introduction

We want to understand what it means for our sentences to have a certain content. If our theory of meaning is to be general, it will have to give an account of the contents of the sentences in which it is formulated. Let's call a theory of meaning that does that "self-sufficient." Such a theory applies to a formulation of itself.

For representationalist theories of meaning, self-sufficiency requires a representationalist account of expressions like "is true" or "refers to." In Tarski's (1944, p. 348) terms, the representationalist needs a "semantically closed language." Hence, the liar paradox is a problem that a representationalist who claims to offer a self-sufficient account of meaning must address. For otherwise the representationalist's theory will be built on a concept of which it offers no account.

---

[1] For helpful discussions and feedback, I would like to thank Lucas Rosenblatt, Andreas Fjellstad, Eduardo Barrio, Luis Estrada-González, Federico Pailos, Diego Tajer, Robert Brandom, Daniel Kaplan, Shuhei Shimamura, Rea Golan, and the audience at talks in Buenos Aires (SADAF), Mexico City (UNAM), Lisbon (LanCog), and, of course, Hejnice, where I presented versions of this material. Thanks to Viviane Fairbank and Mansooreh Kimiagari for proofreading the paper.

Ulf Hlobil

The inferentialist, however, should not point to this splinter in her rival's eye, as there may be a beam in her own eye. For there is an analogue of the liar paradox for inferentialist theories of meaning. That is the topic of this paper. According to inferentialism, what it means for a sentence to have a certain content is that the sentence plays a particular inferential role (Brandom, 1994, 2008; Peregrin, 2014; Sellars, 1953, 1974). We can think of the inferential role of a sentence, S, as two sets of pairs. The first set contains all and only the pairs $\langle X, Y \rangle$ such that the inference from X and S to Y is valid, which I will write as $X, S \vdash Y$. And the second set contains all and only the pairs $\langle X, Y \rangle$ such that $X \vdash S, Y$.[2] So, the inferentialist explains meaning in terms of validity, which I denote by "$\vdash$."[3] Hence, a self-sufficient inferentialist theory of meaning must give an account of expressions like "is valid" or "$\vdash$."

Unfortunately, adding a validity predicate to an object language can easily lead to triviality. My goal in this paper is to contribute to a self-sufficient inferentialism by making some progress on how we can save object language expressions for validity from triviality. In particular, I will present a logic with a validity predicate that captures the meta-inferences of that logic while also allowing us to prove that many arguments are invalid.

The paper is structured as follows. I start with a recap of the debate in Section 2. In Section 3, I present a logic with a validity predicate that captures this logic's meta-inferences and proves for all invalid inferences that don't contain the validity predicate that they are invalid. Section 4 concludes.

2 The Story So Far

In this section, I recapitulate some problems with validity predicates that arise if we allow for self-reference. As self-reference is hard to avoid if one wants to construct a self-sufficient semantic theory (after all, the theory must apply to itself), I will assume that restricting self-reference is off the table. I use "$\overline{A}$" as a canonical name for the sentence A, and "$\overline{\vdash}$" as a canonical

---

[2] I ignore Language Entry and Departure Transitions here (Sellars, 1974, pp. 423–424).

[3] Note that since the inferentialist's notion of validity is wider than logical validity, it will be no help below to say that there is no problem with logical validity (Field, 2017; Ketland, 2012). To see that the relevant notion of validity is broader, notice that the consequence relation that the inferentialist uses is not closed under substitution. It is, e.g., an important part of the meaning of the atomic sentence "a is pink" that the inference from "a is pink" and "b is crimson" to "b is darker than a" is valid. You cannot, however, substitute "a is cramine" for "a is pink" *salva consequentia* here.

name for the set Γ, etc. Thus, I assume self-reference by fiat, even without Gödel-coding.

2.1 The v-Curry Paradox

We want to introduce sentences like $Val(\overline{\Gamma}, \overline{\Delta})$ that express that the inference from Γ to Δ is valid, i.e., that $\Gamma \vdash \Delta$, where Γ and Δ may be sets or sentences. What does it take for a predicate to express validity? Beall and Murzi (2013) suggest that, intuitively, *Val* must obey so-called validity detachment (VD) and validity proof (VP):

$$\frac{}{\Gamma, Val(\overline{\Gamma}, \overline{\Delta}) \Rightarrow \Delta} \text{ VD} \qquad \frac{\Gamma \Rightarrow \Delta}{\Rightarrow Val(\overline{\Gamma}, \overline{\Delta})} \text{ VP}$$

Unfortunately, this yields triviality if we accept contraction and cut. To see this, let κ be (inter-substitutable with) $Val(\overline{\kappa}, \overline{\bot})$, and call substituting κ for $Val(\overline{\kappa}, \overline{\bot})$ or vice versa "κ-substitution." We can now reason thus:

$$\cfrac{\cfrac{\cfrac{\cfrac{\cfrac{\cfrac{}{\kappa, Val(\overline{\kappa}, \overline{\bot}) \Rightarrow \bot} \text{ VD}}{\kappa, \kappa \Rightarrow \bot} \kappa\text{-substitution}}{\kappa \Rightarrow \bot} \text{ contraction}}{\Rightarrow Val(\overline{\kappa}, \overline{\bot})} \text{ VP}}{\Rightarrow \kappa} \kappa\text{-substitution} \quad \cfrac{\cfrac{\cfrac{}{\kappa, Val(\overline{\kappa}, \overline{\bot}) \Rightarrow \bot} \text{ VD}}{\kappa, \kappa \Rightarrow \bot} \kappa\text{-substitution}}{\kappa \Rightarrow \bot} \text{ contraction}}{\Rightarrow \bot} \text{ cut}$$

Non-transitive theorists, like Ripley (2013), respond to such problems by rejecting cut. The most popular non-transitive logic is ST (Cobreros, Egré, Ripley, & van Rooij, 2012, 2013).

2.2 Faithfulness

Barrio, Rosenblatt, and Tajer (2017) have criticized the non-transitive approach by arguing that, whether or not obeying VD and VP is necessary for expressing validity, it is not sufficient. They suggest a further necessary condition, which I call "faithfulness."

Definition 1 (Faithfulness) *A validity predicate, Val, is faithful just in case $Val(\overline{\Gamma_1}, \overline{\Delta_1}), \ldots, Val(\overline{\Gamma_n}, \overline{\Delta_n}) \vdash Val(\overline{\Theta}, \overline{\Lambda})$ is provable iff $\Theta \vdash \Lambda$ follows from $\Gamma_1 \vdash \Delta_1, \ldots, \Gamma_n \vdash \Delta_n$ via a valid meta-inference.*

Barrio et al. (2017) show that if we add a validity predicate to ST in the most obvious way, then the validity predicate is not faithful. I have defended the non-transitive approach against this criticism in earlier work (Hlobil, 2018b). In this paper, I want to expand on my earlier response and address a remaining issue.

Faithfulness can be understood in different ways, depending on what we mean by "valid meta-inference." We could take a meta-inference to be valid (a) if it is an instance of an admissible meta-rule, or (b) if it is an instance of a derivable meta-rule.[4] A meta-rule is admissible in a logic $\mathcal{L}$ iff, for all instances, the conclusion-sequent holds in $\mathcal{L}$ if all the premise-sequents hold in $\mathcal{L}$. A meta-rule is derivable in a sequent calculus iff, for all instances, there is a proof-tree with the conclusion-sequent as its root and all the leaves being either premise-sequents or axioms of the sequent calculus.

We should adopt option (b) and reject (a). That is because if we adopt (a), faithfulness yields triviality if the conditional that we use to define "admissible rule" obeys modus ponens and contraction (in the sense that "if A, then (if A, then B)" is equivalent to "if A, then B").[5] However, modus ponens and contraction are plausible, and the conditional of ST obeys them. Hence, advocates of the non-transitive approach should reject (a).

Someone might be tempted to reject the idea that a predicate that expresses validity must be faithful. That, however, is a bad idea for inferentialists. In formulating and using her semantic theory, the inferentialist is constantly reasoning from premises about validity to conclusion about validity. If the inferentialist rejects the left-to-right direction of faithfulness, then she admits that the inferential role—i.e., the meaning—of "valid" is such that it underwrites inferences that are incorrect, by the standard of what really follows from what. If the inferentialist rejects the right-to-left direction of faithfulness, then she holds that "x is valid; therefore, y is valid" may be correct, by the standard of what actually follows from what, and it may still be ruled invalid, by the meaning of the word "valid," as explained by the inferentialist. Either way, the inferentialist admits that when she describes

---

[4]Dicher and Paoli (2019) suggest a notion of local validity of meta-inferences. This notion is tied to the semantic idea of valuations. I am ignoring it here because of my inferentialist motivation.

[5]I showed this in (Hlobil, 2018a). Fjellstad (ms) has developed the following nice presentation of the point: Let κ be the sentence $Val(\overline{\kappa}, \overline{Val(\Gamma, \Delta)})$, where Γ and Δ are arbitrary. Assume faithfulness for reductio. (i) By faithfulness, $Val(\overline{\kappa}, \overline{Val(\Gamma, \Delta)}) \vdash Val(\overline{\Gamma}, \overline{\Delta})$ is provable if and only if it is the case that if $Val(\overline{\kappa}, \overline{Val(\Gamma, \Delta)}) \vdash Val(\overline{\Gamma}, \overline{\Delta})$, then $\Gamma \vdash \Delta$. (ii) By contraction of the conditional, if $Val(\overline{\kappa}, \overline{Val(\Gamma, \Delta)}) \vdash Val(\overline{\Gamma}, \overline{\Delta})$, then $\Gamma \vdash \Delta$. But this is the right-hand-side of (i) above. (iii) So, by modus ponens, $Val(\overline{\kappa}, \overline{Val(\Gamma, \Delta)}) \vdash Val(\overline{\Gamma}, \overline{\Delta})$. (iv) Since this is the antecedent of (ii), by modus ponens, $\Gamma \vdash \Delta$.

the inferential roles of sentences by using "valid," she is not—by the meaning of this word (as she explains it)—beholden to what actually follows from what. But that amounts to admitting that she is not talking about the actual inferential roles of these sentences. Hence, the inferentialist cannot reject faithfulness.

2.3 Using Contraction Against VD

Unfortunately, another problem arises: contraction, faithfulness, and VD jointly yield triviality. To see this, let κ be $Val(\overline{\kappa}, \overline{Val(\Gamma, \Delta)})$. Now:

$$\cfrac{\cfrac{\cfrac{\cfrac{}{\kappa, Val(\overline{\kappa}, \overline{Val(\Gamma, \Delta)}) \Rightarrow Val(\overline{\Gamma}, \overline{\Delta})} \text{VD}}{Val(\overline{\kappa}, \overline{Val(\Gamma, \Delta)}), Val(\overline{\kappa}, \overline{Val(\Gamma, \Delta)}) \Rightarrow Val(\overline{\Gamma}, \overline{\Delta})} \text{κ-substitution}}{Val(\overline{\kappa}, \overline{Val(\Gamma, \Delta)}) \Rightarrow Val(\overline{\Gamma}, \overline{\Delta})} \text{contraction}}{\kappa \Rightarrow Val(\overline{\Gamma}, \overline{\Delta})} \text{κ-substitution}$$

By faithfulness, the fact that the third line of this tree is provable tells us that, by the application of a derivable meta-rule, we can infer $\Gamma \Rightarrow \Delta$ from $\kappa \Rightarrow Val(\overline{\Gamma}, \overline{\Delta})$. But we have just proven the latter. Hence, there is a way to continue our proof-tree to reach $\Gamma \Rightarrow \Delta$.

We are thus forced to choose between VD and contraction. The option of rejecting contraction is well explored in the literature. I want to look at rejecting VD. But before we worry about VD, we should ask whether we can, in fact, get a faithful validity predicate if we reject VD. Fortunately, the answer is "yes."

2.4 Assuming Sequents

I showed in (Hlobil, 2018b) how to add a faithful validity predicate to ST by rejecting VD. I call this extension of ST "NG," and Fig. 1 gives a variant of it. I use lower case Latin letters and the subscript "0" on sets (e.g., in Ax1) for atomic sentences and sets thereof respectively.

If we dropped the VLR rule from NG, we would have a formulation of ST. And if we also dropped the rules for the truth-predicate, we would have classical propositional logic (with conjunction and the conditional defined in the usual way). Notice that weakening is absorbed into the axioms and that we assume contraction and permutation, i.e., we work with sets on the left and the right.

Axioms of NG

$$\text{Ax1:} \quad \Gamma_0, p \Rightarrow p, \Delta_0$$

Rules of NG

$$\frac{\Gamma \Rightarrow A, \Delta}{\Gamma, \neg A \Rightarrow \Delta} \text{ LN} \qquad \frac{\Gamma, A \Rightarrow \Delta}{\Gamma \Rightarrow \neg A, \Delta} \text{ RN}$$

$$\frac{\Gamma, A \Rightarrow \Delta \quad \Gamma, B \Rightarrow \Delta}{\Gamma, A \vee B \Rightarrow \Delta} \text{ Lv} \qquad \frac{\Gamma \Rightarrow A, B, \Delta}{\Gamma \Rightarrow A \vee B, \Delta} \text{ Rv}$$

$$\frac{\Gamma, A \Rightarrow \Delta}{\Gamma, Tr(\overline{A}) \Rightarrow \Delta} \text{ LT} \qquad \frac{\Gamma \Rightarrow A, \Delta}{\Gamma \Rightarrow Tr(\overline{A}), \Delta} \text{ RT}$$

$$\frac{\overset{1:\ \Gamma_1 \Rightarrow \Delta_1}{\vdots} \quad \cdots \quad \overset{m:\ \Gamma_m \Rightarrow \Delta_m}{\vdots}}{\dfrac{\Theta \Rightarrow \Lambda}{Val(\overline{\Gamma_1}, \overline{\Delta_1}), \ldots, Val(\overline{\Gamma_m}, \overline{\Delta_m}) \Rightarrow Val(\overline{\Theta}, \overline{\Lambda})}} \text{ applications of rules of NG} \\ \text{VLR, [1,\ldots,m]}$$

Figure 1: System NG

What is special about **NG** is that we can assume and discharge sequents by using the VLR-rule. The superscripts that number the assumed sequents are not part of those sequents; they merely help us to keep track of our assumptions. We allow empty discharges.

3 Adding Invalidities to NG

The problem with **NG** that I want to address in this paper is that it cannot prove of any inference that it is invalid. This is a problem because if the inferentialist cannot show of any inference that it is invalid, then, for all we know, the inferential roles of our expressions include any inference you care to name. And while the meanings of our words may be opaque to us in some respects, it is implausible that this opacity is so enormous.

Expressing Validity

3.1 Formulating STV

To address this problem, I want to see how we might extend NG so that we can prove some invalidities. Let's call this extension STV. Ideally the proofs of invalidity in STV should mirror our meta-theoretic knowledge of invalidities in STV. After all, we want to give a treatment, in the object language, of our use of "⊢" in the meta-language. So we should begin by asking ourselves how we usually know about invalidities. For standard sequent calculi, we usually know about invalidities by observing that a root-first proof search fails. My strategy here is to mimic such proof-searches within STV.

At the level of atomic sequents, we perform a root-first proof search simply by seeing whether the sequent has the form of an axiom. To mirror this within STV, let's add the following axioms:

Ax2 If $\Gamma_0 \cap \Delta_0 = \emptyset$ and neither *Val* nor *Tr* occur in $\Gamma_0 \cup \Delta_0$, then $Val(\overline{\Gamma_0}, \overline{\Delta_0}) \Rightarrow$ is an axiom.

With these axioms we can prove of any *Val*-free and *Tr*-free atomic sequent that is not an axiom of ST that it is invalid. After all, for any atomic sequent, $\Gamma_0 \Rightarrow \Delta_0$, that is not of the form of Ax1, Ax2 gives us $Val(\overline{\Gamma_0}, \overline{\Delta_0}) \Rightarrow$, and by RN we get $\Rightarrow \neg Val(\overline{\Gamma_0}, \overline{\Delta_0})$.

Next we observe that the notion of a valid meta-inference used in faithfulness builds in transitivity. We reason transitively in our sequent calculus, and the notion of a valid meta-inference must respect this transitivity of our meta-theoretic reasoning. It does that by relying on the notion of a derivable meta-rule, which cares only about the leaves and the root of a proof-tree and not about the lemmas that must be established in the proof-tree along the way. Let's make this feature of our meta-theoretical reasoning explicit at the object-level of STV by adding the following restricted cut rule.

If the principal operator of every sentence in $\Gamma \cup \Delta \cup \Lambda \cup \Theta$ is *Val* and there are no open assumptions, we can apply the rule:

$$\frac{\Gamma \Rightarrow \Delta, Val(\overline{\Phi}, \overline{\Psi}) \qquad Val(\overline{\Phi}, \overline{\Psi}), \Lambda \Rightarrow \Theta}{\Gamma, \Lambda \Rightarrow \Delta, \Theta} \text{ Val-Cut}$$

We say that a proof-tree of STV is closed iff all undischarged sequents are axioms of STV. A sequent, $\Gamma \Rightarrow \Delta$, is provable in STV (for which we write $\Gamma \vdash_{\text{STV}} \Delta$) iff there is a closed proof-tree of NG that has the sequent as its root.

3.2 Validity and Invalidity in STV

The validity predicate of STV captures all the validities and invalidities of ST. For the validities, it is easy to see that STV captures all of its own validities, which include all ST validities.

Proposition 1 *If* $\Gamma \vdash_{STV} \Delta$, *then* $\vdash_{STV} Val(\overline{\Gamma}, \overline{\Delta})$.

Proof. Suppose that $\Gamma \vdash_{STV} \Delta$ and, hence, $\Gamma \Rightarrow_{STV} \Delta$ is the root of a closed proof-tree. By VLR, $\Rightarrow_{STV} Val(\overline{\Gamma}, \overline{\Delta})$ can be proven in a closed proof-tree. Therefore, $\vdash_{STV} Val(\overline{\Gamma}, \overline{\Delta})$. □

For the invalidities, note that I take the language of ST, $\mathfrak{L}_{ST}$, to not include the validity predicate. The validity predicate of STV captures all ST invalidities in $\mathfrak{L}_{ST}$.

Proposition 2 *If* $\Gamma \not\vdash_{ST} \Delta$ *and* $\Gamma \cup \Delta \in \mathfrak{L}_{ST}$, *then* $\vdash_{STV} \neg Val(\overline{\Gamma}, \overline{\Delta})$.

Proof. Suppose that $\Gamma \not\vdash_{ST} \Delta$ and *Val* does not occur in $\Gamma \cup \Delta$. So there is no proof-tree in ST with $\Gamma \Rightarrow_{ST} \Delta$ as its root. Hence, a root-first proof search (in which we forbid loops and bottom-to-top rule applications) for $\Gamma \Rightarrow_{ST} \Delta$ will result in a tree with at least one leaf $\Theta_0 \Rightarrow_{ST} \Lambda_0$ such that $\Theta_0 \cap \Lambda_0 = \emptyset$ (and hence $\Theta_0 \not\vdash_{ST} \Lambda_0$). Since $\Theta_0 \cap \Lambda_0 = \emptyset$ and *Val* does not occur in $\Theta_0 \cup \Lambda_0$, we have $Val(\overline{\Theta_0}, \overline{\Lambda_0}) \Rightarrow_{STV}$ by Ax2. Since all the rules of ST are invertible, the inverse of the path from $\Theta_0 \Rightarrow_{ST} \Lambda_0$ to $\Gamma \Rightarrow_{ST} \Delta$ is a derivable rule application in ST and, hence, in STV. Call this derivable rule DERIV. We can now reason thus:

$$\cfrac{\cfrac{\cfrac{1: \ \Gamma \Rightarrow \Delta}{\Theta_0 \Rightarrow \Lambda_0} \text{DERIV}}{Val(\overline{\Gamma}, \overline{\Delta}) \Rightarrow Val(\overline{\Theta_0}, \overline{\Lambda_0})} \text{VLR,[1]} \quad Val(\overline{\Theta_0}, \overline{\Lambda_0}) \Rightarrow}{\cfrac{Val(\overline{\Gamma}, \overline{\Delta}) \Rightarrow}{\Rightarrow \neg Val(\overline{\Gamma}, \overline{\Delta})} \text{RN}} \text{Val-Cut}$$

This proof-tree does not have any open assumptions. Therefore, $\vdash_{STV} \neg Val(\overline{\Gamma}, \overline{\Delta})$. □

I will show below that the converses of both propositions also hold and, hence, that the validity predicate of STV (applied to sentences of $\mathfrak{L}_{ST}$) strongly represents ST-validity.[6]

---

[6]To be explicit, $R(\overline{x}, \overline{y})$ strongly represents relation $\mathcal{R}(x, y)$ in STV iff (i) [$\vdash_{STV} R(\overline{x}, \overline{y})$ iff $\mathcal{R}(x, y)$] and (ii) [$\vdash_{STV} \neg R(\overline{x}, \overline{y})$ iff not $\mathcal{R}(x, y)$].

Expressing Validity

Before we turn to that, however, notice that if STV doesn't prove the empty sequent, there must be "validity-gaps," i.e., there must be inferences such that STV proves neither that they are valid nor that they are invalid.

Proposition 3 *If $\emptyset \not\vdash_{\mathsf{STV}} \emptyset$, then there are some sets, Γ, Δ, such that neither $\vdash_{\mathsf{STV}} Val(\overline{\Gamma}, \overline{\Delta})$ nor $\vdash_{\mathsf{STV}} \neg Val(\overline{\Gamma}, \overline{\Delta})$.*

Proof. By example. Let $\kappa = Val(\overline{\kappa}, \overline{\emptyset})$. Suppose we had $\vdash_{\mathsf{STV}} \neg \kappa$ and hence $\kappa \Rightarrow_{\mathsf{STV}}$. By VLR, we would get $\Rightarrow_{\mathsf{STV}} \kappa$. And Val-Cut would then yield the empty sequent. Similarly, if we had $\Rightarrow_{\mathsf{STV}} \kappa$, then faithfulness (which I will prove below) would give us $\kappa \Rightarrow_{\mathsf{STV}}$, and this would yield the empty sequent. □

I will now show that STV doesn't prove the empty sequent. We will first need to establish the following lemma, which will also make it easy to see, along the way, that the validity predicate of STV is faithful.

Lemma 1 *If $Val(\overline{\Gamma_1}, \overline{\Delta_1}), \ldots, Val(\overline{\Gamma_n}, \overline{\Delta_n}) \Rightarrow Val(\overline{\Theta_1}, \overline{\Lambda_1}), \ldots, Val(\overline{\Theta_m}, \overline{\Lambda_m})$ is provable in STV, then, for some $1 \leq i \leq m$, the sequent $Val(\overline{\Gamma_1}, \overline{\Delta_1}), \ldots, Val(\overline{\Gamma_n}, \overline{\Delta_n}) \Rightarrow Val(\overline{\Theta_i}, \overline{\Lambda_i})$ is provable in a proof-tree where the root comes by VLR.*

Proof. We argue by induction on proof-height. Suppose the lemma holds for sequents provable in trees strictly lower than k. A proof of height k of a target sequent must come by VLR, Ax1, a bottom-to-top rule, or Val-Cut.

If it comes by VLR, we are done. If it comes by Ax1, then the left and the right share a sentence. So, for some $1 \leq l \leq n$, $\Gamma_l = \Theta_i$ and $\Delta_l = \Lambda_i$. We assume $\Theta_i \Rightarrow \Lambda_i$ and immediately use VLR, discharging, for all $1 \leq r \leq n$ (vacuously except for $r = l$), the assumptions $\Theta_r \Rightarrow \Lambda_r$. We thus prove $Val(\overline{\Gamma_1}, \overline{\Delta_1}), \ldots, Val(\overline{\Gamma_n}, \overline{\Delta_n}) \Rightarrow Val(\overline{\Theta_i}, \overline{\Lambda_i})$ by VLR.

If the target sequent comes by a bottom-to-top rule application, the premise sequent must include a sentence, A, whose principal connective is not *Val*. The premise of that must come by a top-to-bottom application of the same rule or by Val-Cut. If it comes by the top-to-bottom rule application, we apply our hypothesis to the premise of that step and we are done. If it comes by Val-Cut, the same reasoning applies to the premise of Val-Cut that contains A. The premise comes either by introducing A or by Val-Cut. If the first, we can eliminate this introduction and subsequent elimination. If the second, the same reasoning applies again. The chain of Val-Cut applications must end because the leaves of proof-trees contain only finitely many sentences.

If the root comes by Val-Cut, our two premise-sequents are $Val(\overline{\Gamma_1}, \overline{\Delta_1})$, ..., $Val(\overline{\Gamma_x}, \overline{\Delta_x}) \Rightarrow Val(\overline{\Theta_1}, \overline{\Lambda_1}), \ldots, Val(\overline{\Theta_y}, \overline{\Lambda_y}), Val(\overline{\Xi}, \overline{\Pi})$ and $Val(\overline{\Xi}, \overline{\Pi})$, $Val(\overline{\Gamma_{x+1}}, \overline{\Delta_{x+1}}), \ldots, Val(\overline{\Gamma_n}, \overline{\Delta_n}) \Rightarrow Val(\overline{\Theta_{y+1}}, \overline{\Lambda_{y+1}}), \ldots, Val(\overline{\Theta_m}, \overline{\Lambda_m})$, for some renumbering (and perhaps doubling) of the $\overline{\Gamma}$s, $\overline{\Delta}$s, $\overline{\Theta}$s, and $\overline{\Lambda}$s in the target sequent.

By our induction hypothesis, for both sequents, a sequent like it with all but one sentence on the right deleted can be proven via VLR. Now, if for the first sequent the remaining sentence on the right is not $Val(\overline{\Xi}, \overline{\Pi})$, then we get the desired result by adding some empty discharges. So let's assume that we can prove $Val(\overline{\Gamma_1}, \overline{\Delta_1}), \ldots, Val(\overline{\Gamma_x}, \overline{\Delta_x}) \Rightarrow Val(\overline{\Xi}, \overline{\Pi})$ via VLR. Then there is a proof-tree starting with (some of) the sequents that correspond to the sentences on the left and with a root that corresponds to the sentence on the right. Moreover, from the second premise of our Val-Cut application, we know that, for some s, $(y+1) \leq s \leq m$, and $Val(\overline{\Xi}, \overline{\Pi}), Val(\overline{\Gamma_{x+1}}, \overline{\Delta_{x+1}}), \ldots, Val(\overline{\Gamma_n}, \overline{\Delta_n}) \Rightarrow Val(\overline{\Theta_s}, \overline{\Lambda_s})$ is derivable via VLR. Thus, there is a corresponding proof-tree with the sequents on the left as assumed leaves. We can combine the two proof-trees to get one proof-tree in which $\Theta_s \Rightarrow \Lambda_s$ is derived from the assumptions $\Gamma_1 \Rightarrow \Delta_1, \ldots, \Gamma_n \Rightarrow \Delta_n$, where $\Xi \Rightarrow \Pi$ is the root of a subproof with leaves $\Gamma_1 \Rightarrow \Delta_1, \ldots, \Gamma_x \Rightarrow \Delta_x$. □

With this lemma in hand, it is easy to show that the validity predicate of STV is faithful.

Theorem 1 *The validity predicate of* STV *is faithful, i.e.,* $Val(\overline{\Gamma_1}, \overline{\Delta_1}), \ldots, Val(\overline{\Gamma_n}, \overline{\Delta_n}) \Rightarrow Val(\overline{\Theta}, \overline{\Lambda})$ *is provable in* STV *iff there is a proof-tree in* STV *with* $\Theta \Rightarrow \Lambda$ *as its root and (some of)* $\Gamma_1 \Rightarrow \Delta_1, \ldots, \Gamma_n \Rightarrow \Delta_n$ *as its leaves, i.e., this is an application of a derivable rule.*

Proof. The left-to-right direction is immediate from VLR. The right-to-left direction is immediate from Lemma 1. □

We can now establish that STV does not prove the empty sequent. And as we have already seen above, this implies that STV does not prove all of its own invalidities.

Lemma 2 STV *does not prove the empty sequent, i.e.,* $\emptyset \not\vdash_{\mathsf{STV}} \emptyset$.

Proof. The empty sequent can come only by Val-Cut. The premise-sequents are $\Rightarrow Val(\overline{\Gamma}, \overline{\Delta})$ and $Val(\overline{\Gamma}, \overline{\Delta}) \Rightarrow$. By Lemma 1, the first premise is provable via VLR. Hence, $\Gamma \Rightarrow \Delta$ is provable. However, $Val(\overline{\Gamma}, \overline{\Delta}) \Rightarrow$ is

Expressing Validity

provable only if a root-first proof search for $\Gamma \Rightarrow \Delta$ fails. Thus, there would have be atomic sets Γ_0 and Δ_0 such that $\Gamma_0 \Rightarrow \Delta_0$ and $Val(\overline{\Gamma_0}, \overline{\Delta_0}) \Rightarrow$. The first has to come by Ax1 and the second by Ax2. Because of the former, $\Gamma_0 \cap \Delta_0 \neq \emptyset$. But that means that $Val(\overline{\Gamma_0}, \overline{\Delta_0}) \Rightarrow$ cannot come by Ax2. □

It follows that there is no inference such that STV proves that the inference is valid and also proves that it is invalid.

Corollary 1 *There are no sets Γ and Δ, such that both $\Rightarrow Val(\overline{\Gamma}, \overline{\Delta})$ and $Val(\overline{\Gamma}, \overline{\Delta}) \Rightarrow$.*

As already intimated, however, Lemma 2 together with Proposition 3 also implies:

Proposition 4 *There are some sets, Γ and Δ, such that neither $\vdash_{STV} Val(\overline{\Gamma}, \overline{\Delta})$ nor $\vdash_{STV} \neg Val(\overline{\Gamma}, \overline{\Delta})$.*

In other words, STV doesn't decide all questions about validity. In particular, as we have seen in the proof of Proposition 3, STV shows neither that what the v-Curry sentence says is true, namely that the empty set follows validly from it, nor that it is false.

We have seen above that STV proves all the validities and invalidities of ST. We can now show the converse as well.

Lemma 3 *If $\vdash_{STV} Val(\overline{\Gamma}, \overline{\Delta})$ and $\Gamma \cup \Delta$ is Val-free, then $\Gamma \vdash_{ST} \Delta$.*

Proof. Suppose that $\Rightarrow_{STV} Val(\overline{\Gamma}, \overline{\Delta})$ and $\Gamma \cup \Delta$ is Val-free. Our root must have come by VLR. So, $\Gamma \Rightarrow_{STV} \Delta$ is provable. Since this does not contain Val, it cannot come by VLR or Val-Cut. So it must come by one of the ST-rules. This applies to all the premise sequents and, repeating the reasoning, to all sequents in the proof-tree. Hence, $\Gamma \vdash_{ST} \Delta$. □

Theorem 2 *The STV validity predicate, applied to ST sentences, strongly represents ST validity, i.e., for all $\Gamma \cup \Delta \subseteq \mathcal{L}_{ST}$, we have, first, that $\vdash_{STV} Val(\overline{\Gamma}, \overline{\Delta})$ iff $\Gamma \vdash_{ST} \Delta$; and, second, that $\vdash_{STV} \neg Val(\overline{\Gamma}, \overline{\Delta})$ iff $\Gamma \nvdash_{ST} \Delta$.*

Proof. The right-to-left directions of the two biconditionals are Propositions 1 and 2 above. The left-to-right direction of the first biconditional is Proposition 3. For the left-to-right direction of the second biconditional, we prove the contrapositive. Suppose that $\Gamma \vdash_{ST} \Delta$. We know that this implies $\vdash_{STV} Val(\overline{\Gamma}, \overline{\Delta})$. Assume for reductio that $\vdash_{STV} \neg Val(\overline{\Gamma}, \overline{\Delta})$. That would mean that $Val(\overline{\Gamma}, \overline{\Delta}) \vdash_{STV}$. But this is ruled out by Corollary 1. □

Let's take stock. The validity predicate of STV is faithful, and it captures the validities and invalidities of ST (over a *Val*-free language) perfectly. Moreover, STV doesn't prove of any inference that it is valid and also that it is invalid. However, there are some inferences for which STV proves neither that they are valid nor that they are invalid.

3.3 STV Embeds LP in ST

There is another perspective on the results above that is worth mentioning. It is well-known that LP is the external logic of ST (Barrio, Pailos, & Szmuc, 2019; Barrio, Rosenblatt, & Tajer, 2015; Dicher & Paoli, 2019; Pynko, 2010). That is (Barrio et al., 2015, p. 557):

Fact 1 *Let* t *be a translation function such that* $t(\Gamma \Rightarrow \Delta) = \bigwedge \Gamma \supset \bigvee \Delta$ *and* $t(\Rightarrow \Delta) = \bigvee \Delta$. *Then an* ST*-meta-inference is admissible iff its* t*-translation is* LP*-valid, i.e.,* $\Theta \vdash_{ST} \Lambda$ *whenever* $\Gamma_1 \vdash_{ST} \Delta_1, \ldots, \Gamma_n \vdash_{ST} \Delta_n$ *holds if and only if* $t(\Gamma_1 \Rightarrow \Delta_1), \ldots, t(\Gamma_n \Rightarrow \Delta_n) \vdash_{LP} t(\Theta \Rightarrow \Lambda)$.

So, from an ST-perspective, LP tells us how to reason with claims about validity, i.e., with sequents. The validity predicate of STV allows us to codify this reasoning not at the meta-inferential level of STV but at the inferential level. In particular, for any inference that is LP-valid, we can translate the sentences that occur in it (using the inverse of t) into sequents of ST and, then, into *Val*-sentences of STV. Either the resulting inference is STV-valid or STV proves the negation of one of the premises, i.e.:

Proposition 5 $t(\Gamma_1 \Rightarrow \Delta_1), \ldots, t(\Gamma_n \Rightarrow \Delta_n) \vdash_{LP} t(\Theta \Rightarrow \Lambda)$ *if and only if either* $Val(\overline{\Gamma_1}, \overline{\Delta_1}), \ldots, Val(\overline{\Gamma_n}, \overline{\Delta_n}) \vdash_{STV} Val(\overline{\Theta}, \overline{\Lambda})$ *or, for some* $1 \leq i \leq n$, $\vdash_{STV} \neg Val(\overline{\Gamma_i}, \overline{\Delta_i})$, *where* $\Gamma_1, \ldots, \Gamma_n, \Delta_1, \ldots, \Delta_n, \Theta$, *and* Λ *are Val-free.*

Proof. Left-to-right: Suppose that $t(\Gamma_1 \Rightarrow \Delta_1), \ldots, t(\Gamma_n \Rightarrow \Delta_n) \vdash_{LP} t(\Theta \Rightarrow \Lambda)$. By Fact 1, either one of $\Gamma_1 \vdash_{ST} \Delta_1, \ldots, \Gamma_n \vdash_{ST} \Delta_n$ fails, or $\Theta \vdash_{ST} \Lambda$ holds. If the latter, then $\vdash_{STV} Val(\overline{\Theta}, \overline{\Lambda})$ and we can prove this via VLR. Adding vacuous discharges we can hence also prove: $Val(\overline{\Gamma_1}, \overline{\Delta_1}), \ldots, Val(\overline{\Gamma_n}, \overline{\Delta_n}) \vdash_{STV} Val(\overline{\Theta}, \overline{\Lambda})$. If the former, i.e., $\Gamma_i \not\vdash_{ST} \Delta_i$, then by Proposition 2, $\vdash_{STV} \neg Val(\overline{\Gamma_i}, \overline{\Delta_i})$.

Right-to-left: If, for some $1 \leq i \leq n$, $\vdash_{STV} \neg Val(\overline{\Gamma_i}, \overline{\Delta_i})$, then, by Theorem 2, $\Gamma_i \not\vdash_{ST} \Delta_i$. Hence, $\Theta \vdash_{ST} \Lambda$ whenever $\Gamma_1 \vdash_{ST} \Delta_1, \ldots, \Gamma_n \vdash_{ST} \Delta_n$. So, by Fact 1, $t(\Gamma_1 \Rightarrow \Delta_1), \ldots, t(\Gamma_n \Rightarrow \Delta_n) \vdash_{LP} t(\Theta \Rightarrow \Lambda)$. For the other disjunct, suppose that $Val(\overline{\Gamma_1}, \overline{\Delta_1}), \ldots, Val(\overline{\Gamma_n}, \overline{\Delta_n}) \vdash_{STV} Val(\overline{\Theta}, \overline{\Lambda})$.

Expressing Validity

By faithfulness, $\Theta \Rightarrow \Lambda$ follows in ST via a derivable meta-rule from $\Gamma_1 \Rightarrow \Delta_1, \ldots, \Gamma_n \Rightarrow \Delta_n$. So, whenever we have $\Gamma_1 \vdash_{ST} \Delta_1, \ldots, \Gamma_n \vdash_{ST} \Delta_n$, we also have $\Theta \vdash_{ST} \Lambda$. Therefore, by Fact 1, $t(\Gamma_1 \Rightarrow \Delta_1), \ldots, t(\Gamma_n \Rightarrow \Delta_n) \vdash_{LP} t(\Theta \Rightarrow \Lambda)$. □

This allows us to view LP as a theory that describes how we should reason with sentences about validity. Given that MP fails in LP, the failure of VD in STV can no longer surprise us. The advocate of STV can see LP as merely a disguised formulation of the logic of validity claims.

4 Conclusion

I have presented a way to add to ST a validity predicate that captures the derivable meta-inference of the resulting logic and strongly represents ST-validity over the *Val*-free fragment of the language. Where does that leave us with respect to a self-sufficient inferentialism?

No doubt STV falls short of providing a fully satisfactory account of the meaning of "valid" as this term is used by the inferentialist. It has at least three serious limitations. First, STV is propositional. And since the usual sequent rules for the quantifiers are not invertible, it is not obvious how the method for proving invalidities presented here could be extended to first-order invalidities. Second, STV deems some inferences invalid without proving, in the object language, that they are invalid, e.g., the inference from the v-Curry sentence to an absurd conclusion. This is an aspect of the inferential role of "valid," as used by the inferentialist who puts forward STV, about which her inferentialist account of "*Val*" is silent. Third, in proving things about STV, the inferentialist makes use of powerful mathematical machinery, e.g., in proofs by induction. But there is no analogue of that machinery at the object level of STV.

On the other hand, STV provides a model of what are arguably the two basic ways of knowing about validities and invalidities respectively, given an inferentialist perspective in a sequent calculus setting. We know about validities via proofs, and we know about invalidities via failed root-first proof-searches. STV offers an account of the use of "valid" in these activities. That brings the inferentialist a step closer to an account of her use of "valid" in stating her theory. Further steps will have to wait for another occasion.

References

Barrio, E. A., Pailos, F., & Szmuc, D. (2019). A hierarchy of classical and paraconsistent logics. *Journal of Philosophical Logic*, 1–28. doi: 10.1007/s10992-019-09513-z

Barrio, E. A., Rosenblatt, L., & Tajer, D. (2015). The logics of strict-tolerant logic. *Journal of Philosophical Logic*, *44*(5), 551–571.

Barrio, E. A., Rosenblatt, L., & Tajer, D. (2017). Capturing naive validity in the cut-free approach. *Synthese*(online first), 1–17.

Beall, J., & Murzi, J. (2013). Two flavors of Curry's paradox. *Journal of Philosophy*, *110*(3), 143–165.

Brandom, R. B. (1994). *Making It Explicit: Reasoning, Representing, and Discursive Commitment*. Cambridge, Mass.: Harvard University Press.

Brandom, R. B. (2008). *Between saying and doing: towards an analytic pragmatism*. Oxford: Oxford University Press.

Cobreros, P., Egré, P., Ripley, D., & van Rooij, R. (2012). Tolerant, classical, strict. *Journal of Philosophical Logic*, *41*(2), 347–385.

Cobreros, P., Egré, P., Ripley, D., & van Rooij, R. (2013). Reaching transparent truth. *Mind*, *122*(488), 841–866.

Dicher, B., & Paoli, F. (2019). ST, LP and tolerant metainferences. In C. Bakent & T. Ferguson (Eds.), *Graham Priest on dialetheism and paraconsistency*. Dordrecht: Springer.

Field, H. (2017). Disarming a paradox of validity. *Notre Dame Journal of Formal Logic*, *58*(1), 1–19.

Fjellstad, A. (ms). *Comparing theories definable on Strong Kleene models from within*.

Hlobil, U. (2018a). The cut-free approach and the admissibility-curry. *Thought: A Journal of Philosophy*, *7*(1), 40–48.

Hlobil, U. (2018b). Faithfulness for naive validity. *Synthese, forthcoming*.

Ketland, J. (2012). Validity as a primitive. *Analysis*, *72*(3), 421–430.

Peregrin, J. (2014). *Inferentialism: Why Rules Matter*. New York: Palgrave MacMillan.

Pynko, A. P. (2010). Gentzen's cut-free calculus versus the logic of paradox. *Bulletin of the Section of Logic*, *39*(1/2), 35–42.

Ripley, D. (2013). Paradoxes and failures of cut. *Australasian Journal of Philosophy*, *91*(1), 139–164.

Sellars, W. (1953). Inference and meaning. *Mind*, *62*, 313–338.

Sellars, W. (1974). Meaning as functional classification: a perspective on the relation of syntax to semantics. *Synthese*, *27*, 204–228.

Tarski, A. (1944). The semantic conception of truth and the foundations of semantics. *Philosophy and Phenomenological Research*, *4*(3), 341–376.

Ulf Hlobil
Concordia University
Canada
E-mail: `ulf.hlobil@concordia.ca`

Logic and Ethics

PER MARTIN-LÖF

Abstract:[1] The condition under which it is correct (proper) to make an assertion is that the assertor knows how (is able) to perform the task which constitutes the content of the assertion (correctness condition for assertions). To make an assertion is to commit (obligate) yourself to performing the task which constitutes the content of the assertion (commitment account of assertion). The condition under which it is correct (proper) to undertake an obligation (make a commitment) is that the obligor knows how (is able) to fulfil it (ought implies can). The relation between the preceding three principles is simple: the correctness condition for assertions follows from the commitment account of assertion taken together with the ought-implies-can principle. Both the commitment account of assertion and the ought-implies-can principle bring in the notion of duty (obligation) and hence implicitly, by the correlativity of rights and duties, the notion of right. On the other hand, the notions of right and duty are the key notions of deontological ethics. Thus, all in all, logic has, not only an ontological layer and an epistemological layer, but also a deontological layer underlying the epistemological one. It can be avoided only by treating the notion of knowledge how (can) as a primitive notion, thereby abstaining from relating it to the notions of right and duty (may and must).

Keywords: assertion, request, fulfilment

A more precise formulation of the title would be Logic and Deontology. Let me begin by referring to the structure of a speech act. We have the act of uttering something and that which is uttered, the utterance in that sense, which we in linguistics call a complete sentence:

$$\longrightarrow \text{sentence}$$

---

[1] Per Martin-Löf's invited lecture at Logica 2019 was based on a transcript of a lecture given in March 2019 at the conference *Proof-Theoretic Semantics* in Tübingen. The transcript, which was prepared by Ansten Klev, has previously been published in the online proceedings of that conference, available on the webpages of Tübingen University via the permanent link http://dx.doi.org/10.15496/publikation-35319.

Per Martin-Löf

When you consider a complete sentence, then the outermost structure of it, the most basic structure, is the mood/content structure:

$$\longrightarrow \text{mood content}$$

The view that this is the most basic form of utterance is properly ascribed to Charles Bally, who was a French linguist, successor of Saussure in Geneva: he so to say replaced the subject/predicate form as the most basic form with the mood/content form.

The mood we can characterize at this stage simply as the kind of speech act: if it is an assertion, then we have the assertoric mood, if it is a question, then we have the question mood, if it is a warning, we have the warning mood, etc. But, what about the other part here, the content part? Here I have a proposal. I am going to distinguish between assertoric contents and propositions, or in Dummett's terminology, their ingredient senses. This can be illustrated in the following way, if we have as the mood the assertoric mood, and we take the special case of a content of the form A true, where A is a proposition:

$$\underset{\underset{\text{mood}}{\uparrow}}{\vdash} \underbrace{\overset{\overset{\text{prop}}{\downarrow}}{A} \text{ true}}_{\text{content}}$$

Clearly there is a distinction between content and proposition if you view it in this way. Then we have, on the other hand, the BHK-interpretation of the notion of proposition, where a proposition is identified with an expectation or an intention, in Heyting's terms, or with a task, Aufgabe, in Kolmogorov's terminology. We now have two levels here: we have both propositions and assertoric contents, a proposition being so to say the content made into an object in your theory. And by making it into an object I only mean that we make assertions of the form that something is a proposition: then they are no longer contents, but precisely what we call propositions. On the other hand, there is also a great similarity between propositions and contents, so it seems like a natural idea, if we explain propositions in the way I just referred to, the BHK-way, that we could try that for assertoric contents also. So we would have expectations, or intentions, or tasks, at two levels so to say: on the content level and on the proposition level. This will underlie my talk here from the beginning to the end. So from this point on I look upon the content here as a task in Kolmogorov's terminology.

What is a task then? Simply, something to do, or—in the passive voice—something to be done. As soon as we have a content in this sense we can speak of fulfilling it, or doing it. And then we have immediately also all the temporal modifications and other modifications to which we can subject our verbs.

Now I have fixed the part

$$A \text{ true}$$

What about the mood? Well, for the mood I have already given the general explanation, and then it only remains, for the specific moods that are going to be considered, to give the special explanation for each of them.

Of course, the first mood is the assertoric mood. I will take as the logical notation for assertion the modernized Frege sign, ⊢. To a large extent this talk will be a talk about the explanation of the meaning of ⊢. How is it to be explained? I want to adhere to the point of view that what makes something into an assertion is purely formally that its mood is the assertoric one. If we just prefix the mood to a well-defined content, it is already an assertion, that is, you are going to be held responsible for having made an assertion as soon as you have uttered something with the assertoric mood, just as when you make a promise, you are held responsible for its being a promise however unlikely it may seem that you really are going to fulfil it: it is a promise anyway. Similarly with an assertion: it is an assertion as soon as you have this force sign.

There is a huge literature on the notion of assertion, and it has been made much more accessible by Peter Pagin through his contribution under the entry of 'Assertion' to the Stanford Encyclopedia of Philosophy. There, there is at least half a dozen different views of what assertion is, with endless variations on them, so it makes up a paper of no less than 30 pages or something like that. I am not at all going to contribute to that, I do not have the competence, and it is already available. So I will only take up essentially two views here of what an assertion is, namely the so-called knowledge account of assertion, on the one hand, and on the other hand the commitment account of assertion. I hope to clear up sufficiently how they are related to each other.

Concerning the knowledge account, first of all the term here, to speak of different accounts of assertion, that comes from Williamson. And his own preferred view of the nature of assertion is precisely the knowledge account. But the knowledge account goes back to Frege, we must remember. When Frege defined a judgement as the acknowledgement of the truth of a thought, it was clearly a knowledge account, because of the word 'acknowledgement', which has 'knowledge' in it, and it is the same in the German original: die

Anerkennung der Wahrheit eines Gedankens. So Frege's account of assertion was a knowledge account.

For Frege, the content was the thought, and now I have replaced that by the notion of task, so the question is, What modification does that necessitate as compared with Frege? For Frege it was the acknowledgement of the truth of a thought, so Frege used the word true, or truth. And truth now, when we understand content in the way that I have suggested, corresponds to fulfillability. (Fulfillability works perfectly for expectation and intention, but less well for task, so maybe performability rather, in case you choose task.) So truth corresponds to fulfillability, and then Frege's acknowledgement of the truth of a thought corresponds to acknowledgement of the fulfillability of an intention. And to acknowledge, that is, to get to know, the fulfillability of an intention, that is interpreted in the plainest possible way, namely, that is to grasp how the content is fulfilled. So, to know the thought to be true becomes simply to know how to fulfil the task which makes up the content. That is how the analysis I am giving here is related to Frege's analysis.

Then we may already formulate what it is natural to call the correctness condition for assertion, namely the condition under which it is right, and here several terms are possible to use: right, correct, proper. I am going to use them in the same sense. The condition under which it is right, or correct, or proper, to make an assertion is that you know how to perform the task which constitutes the content of the assertion. This is what I have called the correctness condition for assertion in my abstract.

For acts in general it is usually illuminating to ask, What is the purpose of the act? In this case, if we accept the correctness condition that I just gave, What is the purpose of making an assertion? Then we have already to bring in that the speech act involves not only the speaker, but also the hearer, the receiver of the speech act. So, what is it that the assertor wants to achieve, what is the purpose of making an assertion? Well, if we stick to this knowledge account of assertion that I am discussing right now, then the purpose is nothing but to convey to the hearer that the speaker knows how to fulfil the content, the task which makes up the content. The speech act of assertion has no other purpose than to transmit from the speaker to the hearer the information that the speaker knows how to fulfil the task which makes up the content of the assertion, and this succeeds because the speaker must adhere to the correctness condition for assertion that I just formulated.

Since the speaker is conveying to the hearer that he knows how to do something, he knows how to fulfil this task, that means that this could be useful to the hearer: well, he knows how to do that, which means that I can go to him and get help with doing this, if I am in need of that help. But, as it is now, there is no mechanism for this, because then we have to introduce some more things first. And that brings me to the commitment account of assertion, because what it does is precisely to bring in these extra bits that are needed.

So, now I come to the commitment account of assertion, which has its origin in Peirce's work during a very early stage of the last century, 1902–03, I think. Peirce's view was that an assertion should be understood as a taking on of responsibility, taking responsibility for the content of the assertion. Responsibility and commitment are not significantly different, and commitment, on the other hand, refers to obligation and duty, so we have

>commitment
>obligation
>duty

which means that now the deontic notions have already come in that I referred to in the more precise title.

On the other hand, there is the correlativity of rights and duties, a very fundamental insight due to Bentham right at the beginning of the 1800s— by the correlativity of rights and duties I just mean this, that if I have an obligation, or duty, towards my neighbour, then my neighbour has a right against me, and vice versa. So it is the same action that is carried out, but from my point of view it is an obligation to do it, and from the other person's view it is something that he benefits from by getting me to do it.

So, there is this correlativity of rights and duties, which means that as soon as we have the notions of obligation and duty, we also have the notions of permission, dual to obligation, and right, dual to duty:

commitment	entitlement
obligation	permission
duty	right

Now you see that much more has come into this structure, namely the hearer in addition to the speaker, and these deontic notions and their duals. The duality comes in precisely because of the duality between speaker and hearer. So now I can give a first formulation of the commitment account of assertion, so that it can be compared with what I just said about the knowledge account of assertion. By making an assertion, the speaker assumes the duty of performing the task which constitutes the content of the assertion at the request of the hearer. Now you see more of this duality has come in, because at the end I said 'at the request of the hearer'. So now we have not only the speaker, who is the assertor, but we also have the hearer, who receives the assertion, and now he is going to play an important role here, namely in that he has the right to ask the speaker to fulfil his obligation. So we have now request also coming in here:

speaker	hearer
assertion	request

Now things are beginning to look much more promising, because if we take other speech acts like question and answer, then we take immediately for granted that question and answer have to be explained together: you cannot explain the one without bringing in the other. And if we have a command, for instance, there must also be obeyings of the command: we cannot explain the command without having someone who is commanded and who is obligated to obey the command. So it seems very natural, and strange that it has not become generally accepted, as far as I know, that there is a speech act that is dual to assertion in precisely the same way, namely request. Assertion and request have to be explained together, as we already saw a moment ago in my formulation of the commitment account of assertion.

Now I want to vary that formulation in the same way that I varied the formulation of the knowledge account of assertion, namely by putting it in an explicitly teleological way, by asking, What is the purpose of making an assertion? What I said in other words a moment ago then becomes this: the purpose of an assertion on the part of the speaker is to give the hearer the right to request the speaker to perform the task which makes up the content of the assertion, whereupon the speaker is compelled to fulfil his duty by actually performing the task in question. It is essentially the same content as I gave a few minutes ago, but now formulated in an explicitly teleological way.

If we accept this, then we are in the lucky situation of having discovered a very basic logical structure. Namely, we have first of all the assertion of the speaker, which has this form:

$$\vdash C$$

The speaker makes an assertion, and then the hearer has the right to ask the speaker to fulfil his ability, to put his knowledge-how into practice, and that is a speech act of request:

$$C?$$

Logic and Ethics

When the assertor is requested in this way, he is put under the obligation, or duty, to fulfil C, or to do C. So the conclusion is that C gets done, or C is fulfilled:

$$\frac{\vdash C \quad C?}{\begin{array}{c} C \text{ done} \\ C \text{ fulfilled} \end{array}}$$

This way of writing it makes it look as much as possible like an ordinary inference, but you could also put it, perhaps better, in this way:

$$\frac{\dfrac{\vdash C}{C?}}{C \text{ done}}$$

We have the assertion followed by the request, and then, because all of this has already occurred, we may proceed further to have the speaker to do C.

This is really not one rule, this is a whole scheme of rules: one for each form of assertoric content C. I should give at least one or two examples to elucidate this logical structure. A completely non-logical example—it is logical, but let's speak of it as a non-logical example—is this: you have a child running to its mother saying, Mum, I can swim! That corresponds to the assertion. Then the mother says, Can you? or Show me! (in which case we have an exclamation mark) or something like that, and then as a result of that request, the child actually swims. With this example you already see that this is a practical inference in Aristotle's sense: it is a rule where the conclusion is the performance of an action. Practical syllogism sounds a bit old-fashioned, but practical inference is a perfectly good term that we can use presently. So that is one name for this kind of logical rule. Another possibility is to call it the manifestation rule, or if you think of tests of the kind that we are all engaged in, or examinations, we could call it the examination rule, or test rule.

Knowledge-how, or an ability, is definitely what philosophers call a disposition. Disposition covers a variety of disparate concepts, but at least it is clear that an ability is a disposition. Hence the terminology that has been introduced for dispositions can be used here, in which case the request

$$C?$$

is called the stimulus condition, and the conclusion

$$C \text{ done}$$

is called the manifestation of the disposition. Now, stimulus has a ring that I am not quite happy about, so one could perhaps use prompting condition instead of stimulus condition.

Here is the new logical structure that this talk is basically about. Something now should be said about how it relates to the knowledge account of assertion that I gave previously. Under the knowledge account of assertion I simply stipulated what the condition is for the assertion to be right, namely that the speaker knows how to fulfil the task in question. That is a stipulation: it is right under that provision. But, if we go from that to the teleological account in terms of purpose, then it is no longer stipulated that what gives the speaker the right to make the assertion is that he knows how to fulfil the content: it is no longer stipulated. It must nevertheless still be so, of course, but it requires now an argument why that is needed in order for the purpose to be fulfilled.

One way it is simple, namely that if the speaker knows how to fulfil the content, the task, then this interaction works properly, because if he knows how to do it and gets challenged, then he simply does it, and it is no problem for him to do that, because he can do it. It is sufficient that he knows how to fulfil that task, but in the other direction, that it is also necessary, you need to invoke the ought-implies-can principle, as I said in my abstract. Because, if he makes the assertion

$$\vdash C$$

then by so doing, he is undertaking a conditional obligation, namely the conditional obligation

$$\frac{C?}{C \text{ done}}$$

And by the ought-implies-can principle, in order to have the right to undertake an obligation, you must be able to fulfil it. Since you are assuming an obligation, you must be able to fulfil it, and that is precisely the condition that we need for this. So it is both necessary and sufficient that the speaker knows how to fulfil the task that makes up the content.

If we look at the rule

$$\frac{\vdash C \quad C?}{C \text{ done}}$$

you see that the major premiss here is connected with *can*, because the speaker must know how to fulfil the intention—know-how and can I make no difference between. Then we have the hearer, he gets the right to challenge the assertion: he gets the right, which means that he *may* challenge the

Logic and Ethics

assertor. And when the assertor has been challenged in this way, he is under an obligation, so he *must* do something:

$$\frac{\text{can} \quad \text{may}}{\text{must}}$$

I have put it this way just to make it plausible that this is a natural analysis, because can, may and must are among the auxiliary verbs, the main modal auxiliaries, and it seems quite natural that they come in a package, so to say: they fit together into this pattern, and you cannot explain one of them without also bringing in the other two.

Dummett proposed to lift the introduction and elimination pattern from its usual place due to Gentzen, to shift it to the level of assertions, or even to utterances in general, since I began with utterances in general. So he distinguished between conditions for an utterance and consequences of an utterance: what follows from an utterance as compared with what the utterance follows from. Now we have something like this, because an ordinary inference has assertions as premisses and an assertion as conclusion:

$$\frac{\vdash C_1 \ \ldots \ \vdash C_n}{\vdash C} \qquad (C\text{-intro})$$

And this we can consider now as an introduction rule for the form of assertoric content, C, that you have in the conclusion. We now have also the dual rule here, namely

$$\frac{\vdash C \quad C?}{C \text{ done}} \qquad (C\text{-elim})$$

This clearly then should be considered as an elimination rule, since $\vdash C$ occurs as major premiss, an elimination rule for the form of content that C has. So you have an introduction and elimination pattern here arising on the level of assertions.

That brings me to my final remark. I began by saying that this whole lecture will be roughly about what the meaning is of the assertion sign. We are used to the fact that when we ask for the meaning of some linguistic construction, it should be visible somehow from the rules that govern that construction, in general Wittgensteinian terms. The first example of this is of course Gentzen's suggestion that the logical operations are defined by their introduction rules. What about the assertion sign? If you did not have this new rule (C-elim), you would only have the usual rules of inference, which are of the form (C-intro). If you were to take the assertion sign to be determined by these rules, the assertion $\vdash C$ could not mean anything other than that C has been demonstrated, has been inferred by the usual inference rules. And that is not how Frege introduced the assertion sign, what Frege meant by the assertion sign. I explained that earlier on: it is the acknowledgement of the truth of a content that the assertion sign expresses. So, we simply cannot explain the assertion sign by referring to the rules governing it if you only have the rules (C-intro). But now we are in a better situation, because we also have the rules (C-elim), and they are precisely the rules that are meaning-determining for the assertion sign.

Per Martin-Löf
Stockholm University
Sweden
E-mail: `pml@math.su.se`

Revision Operator Semantics: Some Order for the Zoo of Conditionals

ERIC RAIDL[1]

Abstract: This paper introduces a new semantics for $\neg, \wedge,$ and a conditional $>$, where each logical operator is based on a different revision operator plus an additional relation. The conditional is based on a generalised Ramsey Test, negation generalises modal and conditional negation, and conjunction generalises dynamic and intensional conjunction. It is shown that classical logic, non-monotonic conditional logics, intuitionistic and relevance logic, the strict conditional in modal logic, and the logic of update and truth-maker semantics are all special cases and can be obtained by tuning the semantic parameters.

Keywords: conditional logics, Ramsey Test, strict conditional, variably-strict conditional intuitionistic logic, relevance logic, dynamic semantics, update semantics, truthmaker semantics

1 Introduction

More than one century of work on conditionals—sentences of the form "if A, [then] C"—gave rise to a zoo of conditional logics and semantics. These include classical logic with the material implication, modal logic with strict implication (MacColl, 1908), and normal and classical conditional logics (Chellas, 1975) with variably strict conditionals (Lewis, 1973; Stalnaker, 1968). Whereas the latter deviate from classical logic only by conceiving of the conditional as a modal, other accounts also modalise other connectives. A modal negation appears in intuitionistic logic, relevance logic (Routley, Meyer, Plumwood, & Brady, 1983; Urquhart, 1972), or Hype (Leitgeb, 2019). A modal conjunction arises in update semantics (Gillies, 2004; Veltman, 1996) and in truthmaker semantics (Fine, 2012).

This paper develops a unifying semantics where each logical operator is based on a different revision operator, plus a relation. It is shown how

---

[1] Funding for this research was provided by the Deutsche Forschungsgemeinschaft (DFG, German Research Foundation), Research Unit FOR 1614 and under Germany's Excellence Strategy – EXC-Number 2064/1 – Project number 390727645.

some of the above logics or semantics arise as special cases, by tuning the semantic parameters. By this comparison, the differences between the various accounts are reduced to variations of the Ramsey Test and negation or conjunction being modal or not. §2 introduces the revision operator semantics, §3 compares it to other known semantics and §4 proves results about the relation between the former and the latter.

2 Revision Operator Semantics

The language $\mathcal{L}$ is generated from propositional variables Var, negation $\neg$, conjunction $\wedge$, and a conditional $>$. The language with necessity $\Box$ is $\mathcal{L}_\Box$. Dropping a connective c is denoted $\mathcal{L}_{\setminus\{c\}}$. $f\colon X \longrightarrow Y$ is a total function.

Definition 1 $\mathfrak{F} = \langle\, S, R^\neg, R^\wedge, R^>, L, M, E\,\rangle$ *is a* revision frame[2] *for* $\mathcal{L}$ *iff*

1. $S \neq \emptyset$, *the* states.
2. $R^x\colon S \times \mathcal{L} \longrightarrow \wp(S)$ *for* $x \in \{\neg, \wedge, >\}$, *the* revision operators.
3. $L \subseteq S \times \mathcal{L}$, *the* lacking *operator*.
4. $E \subseteq S \times \mathcal{L}$, *the* endorsing *operator*.
5. $M \subseteq S \times S$, *the* matching *operator*.

$\mathfrak{M} = \langle\,\mathfrak{F}, V\,\rangle$ *is a* revision model *for the language* $\mathcal{L}$ *iff* $\mathfrak{F}$ *is a revision frame and* $V\colon S \longrightarrow \wp(\mathrm{Var})$ *is the* valuation.

The revision operators R^x might differ depending on their mode, indicated by $x \in \{\neg, \wedge, >\}$. Given a state $s \in S$ and a sentence $\varphi \in \mathcal{L}$, $R^x(s, \varphi)$ selects admissible φ-revisions of s according to the mode x. Thinking of $R^x(., \varphi)$ as a relation, I write $sR^x_\varphi s'$ iff $s' \in R^x(s, \varphi)$, and $R^x_\varphi(s) = \{s' \in S : sR^x_\varphi s'\}$. $sL\varphi$ means that state s lacks φ. $sE\varphi$ means that s endorses φ, and $s'Ms$ means that s' matches s. The classical[3] interpretation of lacking is non-satisfaction (L is $\nvDash$). Another interpretation is that φ is falsified or disbelieved, or that a φ-lacking state is φ-absurd, or simply absurd. The classical

---

[2]These are not 'proper frames'. A *proper revision frame* would be a structure, where R^x, L, M and E are not formulated with respect to the language, but with respect to the set of propositions. For this one replaces $\mathcal{L}$ by $\wp(S)$ in the defining clauses. For all but one example (update semantics), we could take proper frames, replacing $\mathcal{L}$ by $\wp(S)$ and modifying Definition 2 accordingly (see footnote 5).

[3]The choice of the term "classical" will become clear in §3.

interpretation of endorsement is satisfaction ⊨, but for some purposes it will be interpreted as belief. The classical interpretation of matching is that the relation M contains the identity relation. The rational is that at least identical states match. Other interpretations of M are resemblance, similarity or equivalence. For truthmaker semantics, we need to relax $R^>$ to be a partial function[4] and, for intensional conjunction, to extend M to a three place relation $M \subseteq S^3$, where $s's''Ms$ means that s is the result of fusing s' with s''. All this becomes clearer once we consider the truth clauses and compare them to clauses from known semantics:

Definition 2 Satisfaction *in a state in a revision model is defined as:*

1. $s \vDash p$ *iff* $p \in V(s)$.
2. $s \vDash \neg\varphi$ *iff for all* $s' \in R^\neg(s, \varphi)$, $s'L\varphi$.
3. $s \vDash \varphi \wedge \psi$ *iff there are* s', s'', *such that* $s' \in R^\wedge(s, \varphi)$, $s'' \in R^\wedge(s', \psi)$ *and* $s''Ms$ *[$s's''Ms$ for the ternary M]*.
4. $s \vDash \varphi > \psi$ *iff for all* $s' \in R^>(s, \varphi)$, $s'E\psi$.

I denote $[\varphi] = \{s \in S : s \vDash \varphi\}$.[5] R^x is *intensional* iff $[\varphi] = [\psi]$ implies $R(s, \varphi) = R(s, \psi)$. Similarly for L, E.[6] I use 'satisfaction' instead of truth, to allow for a flexible interpretation of ⊨. I.e., ⊨ can mean truth, but needs not. It can also mean verification, support, acceptance, belief, knowledge, or something else. Thus the states $s \in S$ need not be understood as worldly states. They can also be belief states or information states, as in update semantics. They can be situations, as in situation semantics, or states as in state-semantics for relevance or for truthmaker semantics. Thus a state can be incomplete, i.e., it is not the case that for every sentence φ, $s \vDash \varphi$ or $s \vDash \neg\varphi$ (a *gap*). A state can also be inconsistent, i.e., overdetermined, so that we can have both $s \vDash \varphi$ and $s \vDash \neg\varphi$ (a *glut*). Whenever a state is a world state (no gaps, no gluts), I write it w.

3 Conditionals, Negations and Conjunctions

This section explains each satisfaction clause, comparing it to clauses from known semantics. §4 generalises this by proving point-equivalences.

---

[4] Similarly for Hype.
[5] For a 'proper' revision model, i.e., based on a proper frame (see footnote 2), φ and ψ appearing in the defining clauses for the logical operators need to be replaced by $[\varphi]$ and $[\psi]$.
[6] Models with all operators being intensional can be based on proper frames.

3.1 Conditional: $s \vDash \varphi > \psi$ iff $\forall s' \in R^>(s, \varphi), s' E \psi$.

The conditional clause generalises the *Ramsey Test* (Ramsey, 1929, 1990): The conditional $\varphi > \psi$ holds (is accepted) in state s iff revising by φ leads to believing ψ. My generalisation is threefold: first, several *admissible* revision*s* are considered as opposed to only one. Hence the new test is *imprecise*, as opposed to *precise*. Second, instead of imposing *belief* of the consequent, the defining clause requires *endorsement* of the consequent. Endorsement is my catch-all term for ways of 'holding', which can mean truth, belief, acceptance or something different. Third, instead of conceiving of $R^>(s, \varphi)$ as a (belief) revision of the (belief) state s by the sentence φ, I think of it as the operation of *hypothesising* φ in the state s. Hypothesising is again a catch-all term for different operations, among which revision of beliefs or other doxastic states, but also other doxastic operations such as supposing, updating or conditionalisation. The term also includes ontic operations such as jumping to accessible worlds, to closest or most similar worlds, or adding a φ-state to s and producing thereby one or several new states (e.g., by fusion or by truthmaker-transition).

The generalised Ramsey Test under consideration is thus:

HYPOTHETICAL TEST: $\varphi > \psi$ holds in state s iff *hypothesising* φ in s yields scenario states s' in which ψ is *endorsed*.

Different semantics for the conditional can then be seen as different ways of spelling out 'hypothesising' and 'endorsing'. They might also differ depending on whether hypothesising (and the test) is precise, or imprecise; or depending on whether $R^>$ is *primitive*, i.e., non-reducible to any other operation(s), or *derived*, i.e., $R^>$ is defined by more fundamental operations. Let me show how different semantics implement my Hypothetical Test.

In classical logic, the *material implication* is defined in world states by

- $w \vDash \varphi \to \psi$ iff if $w \vDash \varphi$ then $w \vDash \psi$ [equivalently: $w \nvDash \varphi$ or $w \vDash \psi$].

This can be simulated in revision operator semantics by imposing that E is satisfaction, and $R^>(s, \varphi) = \{s\}$ if $s \vDash \varphi$ and else $= \emptyset$.[7] Call this the *classical interpretation* of E and $R^>$. Classical hypothesising is minimally cognitively demanding, since we keep s, if φ holds, or else we erase s.

In standard models (Chellas, 1975), a generalisation of Stalnaker's semantics, the *variably strict conditional* is defined by

---

[7]Equivalently: $R^>(s, \varphi) := \{s' : s' = s, s' \vDash \varphi\}$.

Revision Operator Semantics

- $w \vDash \varphi \,\square\!\!\rightarrow\, \psi$ iff for all $w' \in R(w, \varphi)$, $w' \vDash \psi$.

Additionally, $R(s, \varphi)$ is intensional. Thus we can think of the selection R as being of the new type $R\colon W \times \wp(W) \longrightarrow \wp(W)$. The variably strict conditional is then obtained with classical endorsement (E is $\vDash$) and an intensional $R^>$. Standard hypothesising is more demanding than classical hypothesising, since it involves considering closest φ-worlds.

The *strict implication* in Kripke models is defined in world states as:

- $w \vDash \square(\varphi \to \psi)$ iff for all $w' \in R(w)$ if $w' \vDash \varphi$ then $w' \vDash \psi$.

where $R \subseteq W^2$. We can simulate this, with E classical ($\vDash$) and setting $R^>(w, \varphi) := R(w) \cap [\varphi]$. This yields an intensional $R^>$, reducible to the accessibility relation R and φ-satisfaction. The difference between strict hypothesising and the previous standard hypothesising is that one considers accessible φ-worlds instead of closest φ-worlds.

In semi-lattice semantics (Urquhart, 1972), the *relevance conditional* is

- $s \vDash \varphi \rightharpoonup \psi$ iff for all t such that $t \vDash \varphi$, $s \circ t \vDash \psi$.

Here $s \circ t$ is the fusion of s with t. We can simulate this with E classical and $R^>(s, \varphi) = \{s \circ t \in S : t \in S, t \vDash \varphi\}$. A similar idea appears in Hype (Leitgeb, 2019), where $\circ$ is partial. These 'additive' versions of hypothesising add a φ-state to the initial state and consider the result, supposed unique, instead of just jumping to closest or accessible φ-states.

In all previous examples E is always classical (satisfaction). An example of a non-classical E appears in truthmaker semantics, where Fine (2012) proposes to define (simple) counterfactual conditionals in world states w, by reference to non-worldly states:

- $w \vDash \varphi \rightarrow \psi$ iff for all t and u, such that $t \to_w u$ and $t \vDash \varphi$, there is $u' \sqsubseteq u$ such that $u' \vDash \psi$.

$\to_w$ is a transition relation and $\sqsubseteq$ is the part relation. The idea is that the conditional is verified (i.e., true) in w iff adding a φ-state t to w results in a state u which contains a ψ-state. We can simulate this with $uE\psi$ iff $\exists u' \sqsubseteq u, u' \vDash \psi$, and the partial $R^>(w, \varphi) = \{u \in S : \exists t, t \vDash \varphi, t \to_w u\}$. The non-classical endorsing is an implicit Kripke possibility and $R^>$ is derived from transition and satisfaction. This is maybe cognitively the most demanding way of hypothesising, since not only do we add a φ-state to the initial state, this addition may also result in different outcomes.

3.2 Negation: $s \vDash \neg\varphi$ iff $\forall s' \in R^{\neg}(s, \varphi), s' L \varphi$.

Classical negation

- $w \vDash \neg\varphi$ iff $w \nvDash \varphi$,

is obtained with $R^{\neg}(s, \varphi) = \{s\}$ and lacking being non-satisfaction (L is $\nvDash$). Call this the *classical interpretation* of $R^{\neg}$ and of L. In general, negation in revision operator semantics generalises two types of negation—'modal negation' (Berto, 2015) and 'conditional negation'.

MODAL NEGATION: Negation is implicitly modal, namely $\neg\varphi$ iff "$\Box \neg_c \varphi$", where the inner negation $\neg_c$ is conceived as being classical.

In Kripke semantics modal negation looks like this:

- $s \vDash \neg\varphi$ iff $\forall s' \in R(s), s' \nvDash \varphi$.

For example, *relevance negation* has this form:

- $s \vDash \neg\varphi$ iff $s^* \nvDash \varphi$,

with $*$ the Routley-star operator. And *Hype negation* (Leitgeb, 2019):

- $s \vDash \neg\varphi$ iff $\forall s'$, if $s \not\perp s'$, then $s' \nvDash \varphi$. (s-compatible states dissatisfy φ)

With L classical ($\nvDash$), classical negation uses $R^{\neg}(s, \varphi) = \{s\}$, relevance negation $R^{\neg}(s, \varphi) = \{s^*\}$, Kripke-modal negation $R^{\neg}(s, \varphi) = R(s)$ and Hype negation $R^{\neg}(s, \varphi) = \{s' \in S : s \not\perp s'\}$. Whereas classical and relevance negation are based on a precise revision—$R^{>}(s, \varphi)$ is a singleton—Kripke-modal and Hype negation are based on an imprecise revision—$R^{>}(s, \varphi)$ is a set.

Ripley's (2012) *weak negation*, defined in neighbourhood semantics, is a modal negation with non-classical L:

- $s \vDash \neg\varphi$ iff $[\varphi] \notin N(s)$ (i.e., $[\varphi]$ is not in the s-neighbourhood).

Another example is *truthmaker negation* (Fine, 2012):

- $s \Vdash \neg\varphi$ iff $s \dashv \varphi$ (i.e., $\neg\varphi$ is verified iff φ is falsified).

Now $R^{\neg}$ is classical ($R^{\neg}(s, \varphi) = \{s\}$). Neighbourhood negation is obtained with $sL\varphi$ iff $[\varphi] \notin N(s)$, and truthmaker negation with L falsification ($sL\varphi$ iff $s \dashv \varphi$). Revision operator negation generalises both to a non-classical $R^{\neg}$. 'Falsification' of φ is not (only) tested in s but in all (φ)-compatible states to s. In brief, contrary to all modal negations, $R^{\neg}$ can be sentence dependent for revision operator negation. This brings me to:

Revision Operator Semantics

CONDITIONAL NEGATION: Negation $\neg\varphi$ is a conditional "$\varphi \triangleright \bot$".[8]

$\varphi \triangleright \bot$ is sometimes called "outer impossibility" of φ. We obtain conditional negation with $L\varphi$ iff $L\bot$. The implicit conditional $\triangleright$ is based on the sentence dependent selection $R^\neg$: $\forall s' \in R^\neg(s,\varphi), s'L\bot$.

Conditional negation is known from Kripke's *intuitionistic negation*, explicitly defined by

- $s \vDash \neg\varphi$ iff $s \vDash \Box(\varphi \to \bot)$ iff for all $s' \in R(w)$, if $s' \vDash \varphi$ then $s' \vDash \bot$.

Explicitly definable negations ($\triangleright$ is $>$) can be obtained, in revision operator semantics, by setting $R^\neg = R^>$ and $s'L\varphi$ iff $s'E\bot$. Then

$s \vDash \neg\varphi$ iff $\forall s' \in R^\neg(s,\varphi), s'L\varphi$ iff $\forall s' \in R^>(s,\varphi), s'E\bot$ iff $s \vDash \varphi > \bot$.

When E is classical ($\vDash$), lacking just means "absurd". This procedure applies to classical, intuitionistic, and update negation (see later).

Negation in revision operator semantics is more general than intuitionistic negation. First, we can model an *implicitly* conditional negation, i.e., based on a conditional not in the language. In that case, $R^\neg \neq R^>$. For example, in Hype, we have $s \vDash \bot$ for no state. Hence *Hype negation* can be rephrased as

- $s \vDash \neg\varphi$ iff $\forall s' \in R^\neg(s,\varphi), s' \vDash \bot$

where $R^\neg(s,\varphi) = \{s' : s \not\sqsubseteq s', s' \vDash \varphi\}$. Thus Hype negation is based on an implicit strict conditional. Revision operator negation generalises this by requiring $s'L\varphi$ instead of $s' \vDash \bot$.

A reason for some negations to be both, modal and conditional, is their interdefinability in Kripke semantics, when $s' \vDash \bot$ for no state. Then the two following are equivalent:

- $\forall s' \in R(s), s' \nvDash \varphi$,
- $\forall s' \in R(s)$ if $s' \vDash \varphi$ then $s' \vDash \bot$.

Without the assumption, the first implies the second, but not vice versa.

A conditional negation not rephrasable as modal negation is *update negation* (Gillies, 2004; Veltman, 1996), where one can show:

- $s \vDash \neg\varphi$ iff $s[\varphi] \vDash \bot$ (i.e., the φ-update of s is absurd).

This can be simulated in revision operator semantics with $R^\neg(s,\varphi) = \{s[\varphi]\}$ and $s'L\varphi$ iff $s' \vDash \bot$.

---

[8] For our purpose, $s \vDash \bot$ only in absurd states, if any. Absurd states have $a \vDash \psi$ for all ψ.

3.3 Conjunction: $s \vDash \varphi \wedge \psi$ iff $\exists s' \in R^{\wedge}(s, \varphi) \exists s'' \in R^{\wedge}(s', \psi), s'' M s$.

Classical conjunction, also called extensional conjunction:

- $s \vDash \varphi \wedge \psi$ iff $s \vDash \varphi$ and $s \vDash \psi$,

is obtained by taking $R^{\wedge}(s, \varphi) = \{s\}$ if $s \vDash \varphi$ and else $= \emptyset$ and conceiving of M as the identity $=$.[9] Call this *classical interpretation* of $R^{\wedge}$ and M. (Note: the classical $R^{\wedge}$ is like the classical $R^{>}$.)

Some semantics have non-classical conjunctions. Consider *update conjunction*:[10]

- $s \vDash \varphi \wedge \psi$ iff $s[\varphi][\psi] = s$ (updating s by φ, then by ψ, returns s).

This can be captured in the revision operator semantics if we take M to be classical (the identity) and $R^{\wedge}(s, \varphi) = \{s[\varphi]\}$. Revision is precise. Revision operator semantics generalises this idea to imprecise revisions and by relaxing identity to 'matching'.

For classical and update conjunction, we could have used the universal instead of the existential quantifier. In fact, update conjunction (in $\mathcal{L}$) reduces to classical conjunction (shown later). This seizes to be so, if we add a modal such as 'it is known'. The more substantial reason for our existential conjunction is *intensional conjunction* (Mares, 2012):

- $s \vDash \varphi \wedge \psi$ iff there are t and u such that $t \vDash \varphi$, $u \vDash \psi$ and $Rtus$

with $R \subseteq S^3$. To simulate this, we extend M to a three place relation $s's'' M s$. Then intensional conjunction is obtained with $R^{\wedge}(s, \varphi) = [\varphi]$, and $tuMs$ iff $Rtus$. In truthmaker semantics $tuMs$ means $t \sqcup u = s$, i.e., s is the fusion of t and u.

In brief, revision operator semantics generalises (1) the Ramsey Test conditional, (2) modal and conditional negation, as well as (3) update and intensional conjunction. Classical connectives can always be obtained.

4 Point-equivalences

In this section I relate different known semantics to the revision operator semantics by showing point-equivalence results.

I will use an adapted notion of point-equivalence.

---

[9]It suffices to assume that the relation M contains the identity relation.
[10]Veltman (1996) and Gillies (2004).

Revision Operator Semantics

Definition 3 *Let $\mathfrak{M} = \langle S, \ldots \rangle$ and $\mathfrak{N} = \langle X, \ldots \rangle$ be models. $\mathfrak{M}$ is point equivalent to $\mathfrak{N}$ (in $\mathcal{L}$) iff there is $f \colon S \longrightarrow X$, such that for each $\varphi \in \mathcal{L}$ and each $s \in S$, we have $s \vDash_{\mathfrak{M}} \varphi$ iff $f(s) \vDash_{\mathfrak{N}} \varphi$.*

The following should be clear:

Theorem 1 *Let $\mathbf{M}$ and $\mathbf{N}$ be classes of models. If every $\mathfrak{M} \in \mathbf{M}$ is point equivalent to some $\mathfrak{N} \in \mathbf{N}$ (denote it $\mathbf{M} \Rightarrow \mathbf{N}$), then $\mathbf{N} \vDash \varphi$ implies $\mathbf{M} \vDash \varphi$.*

As a consequence, if $\mathbf{M} \Leftrightarrow \mathbf{N}$, both classes have the same logic. Furthermore, assume $\mathbf{M} \Rightarrow \mathbf{N}$, then L is sound for $\mathbf{M}$, provided it is sound for $\mathbf{N}$; and L is complete for $\mathbf{N}$, provided it is complete for $\mathbf{M}$.

Example 1 (Classical Logic) The *classical model* T is the set of truth functions $t \colon \text{Var} \longrightarrow \{0, 1\}$ with $\neg, \wedge, >$ classical.

We know how to obtain the material conditional, classical negation and classical conjunction in revision operator semantics: $R^{\neg}$ is the identity; $R^{>}(s, \varphi) = R^{\wedge}(s, \varphi) := \{s\}$ if $s \vDash \varphi$ and else $= \emptyset$; L is $\nvDash$; E is $\vDash$ and M is the identity. From now on, we drop classical operators or relations, when writing down the revision model (although they are implicitly there). Thus the revision model with $\neg, \wedge$ and $>$ classical reduces to $\langle S, V \rangle$—a *classical revision model*.

Theorem 2 *The classical model T is point equivalent to a classical revision model. And any classical revision model is point equivalent to T.*

Proof. Given the classical model T, we defined $\mathfrak{M} = \langle S, V \rangle$ with $S = T$ and $V(t) = \{p \in \text{Var} : t(p) = 1\}$. Thus we have $t \vDash p$ iff $p \in V(t)$ iff $t(p) = 1$. To show the remaining point equivalence, it suffices to show that $\neg, \wedge, >$ are classical, which we already know. Conversely, consider a classical revision model $\langle S, V \rangle$ and map $s \in S$ to the truth function $t_s(p) = 1$ iff $p \in V(s)$. Point equivalence is shown in the same manner. $\square$

Let us continue with examples where $\neg$ and $\wedge$ are classical, but $>$ is not.

Example 2 (Strict implication) $\mathfrak{M}$ is a *Kripke model* iff $\mathfrak{M} = \langle W, R, V \rangle$, where $W \neq \emptyset$, $R \subseteq W^2$ and $V \colon W \longrightarrow \wp(\text{Var})$. We write $R(w) = \{v \in W : \langle w, v \rangle \in R\}$. The truth clauses are classical for $\neg, \wedge$ and:

- $w \vDash p$ iff $p \in V(w)$
- $w \vDash \Box \varphi$ iff for all $v \in R(w)$, $v \vDash \varphi$.

In a Kripke model, we can introduce the material implication $\varphi \to \psi \equiv \neg(\varphi \wedge \neg \psi)$ and strict implication $\varphi > \psi \equiv \Box(\varphi \to \psi)$. A *Kripke revision model* is of the form $\langle S, R^{>}, V \rangle$, where $R^{>}_{\top}$ is a relation over S and

101

$R_\varphi^>(s) = R_\top^>(s) \cap [\varphi]$. Recall: dropping $R^\neg, R^\wedge, E, L$, or M means that they are classical, in particular E is $\vDash$. Furthermore, we define $\Box\varphi \equiv \top > \varphi$.

Theorem 3 *Over $\mathcal{L}_\Box$, any Kripke model $\mathfrak{M}$ is point-equivalent to a Kripke revision model $\mathfrak{M}^* = \langle W, R^>, V \rangle$. And conversely.*

Proof. $\Rightarrow$ direction: set $R_\top^> := R$. $\Leftarrow$ direction: set $R := R_\top^>$. Point-equivalence is shown inductively as follows: $p, \neg, \wedge$ are clear, by classicality. For $>$ recall that $\varphi \to \psi$ abbreviates $\neg(\varphi \wedge \neg\psi)$:

$$w \vDash_\mathfrak{M} \Box(\varphi \to \psi) \quad \text{iff} \quad \forall w' \in R(w), \text{ if } w' \vDash_\mathfrak{M} \varphi, \text{ then } w' \vDash_\mathfrak{M} \psi \quad \text{(Def)}$$
$$\text{iff} \quad \forall w' \in R(w), \text{ if } w' \vDash_{\mathfrak{M}^*} \varphi, \text{ then } w' \vDash_{\mathfrak{M}^*} \psi \quad \text{(IH)}$$
$$\text{iff} \quad \forall w' \in R^>(w, \varphi), w' E \psi \quad (R^>, E)$$
$$\text{iff} \quad w \vDash_{\mathfrak{M}^*} \varphi > \psi \quad (>)$$

$w \vDash_\mathfrak{M} \Box\varphi$ iff $w \vDash_\mathfrak{M} \Box(\top \to \varphi)$ iff (IH) $w \vDash_{\mathfrak{M}^*} \top > \varphi$ iff $w \vDash_{\mathfrak{M}^*} \Box\varphi$. □

Thus, the conditional is the strict implication, if $\neg, \wedge$ are classical, E is satisfaction, $R_\top^>$ is a Kripke relation and $R_\varphi^>(w) = R_\top^>(w) \cap [\varphi]$.

Another paradigmatic example, where $\neg$ and $\wedge$ are classical, but $>$ is not, are variably strict conditionals as modelled by selection functions. The most general such selection function selects a set of sentences.

Example 3 (variably strict implication) $\mathfrak{M}$ is a *sentence selection model* iff $\mathfrak{M} = \langle W, F, V \rangle$ where $W \neq \emptyset$, $F: W \times \mathcal{L} \longrightarrow \wp(\mathcal{L})$ and $V: W \longrightarrow \wp(\text{Var})$. The truth clauses are as in a Kripke model for $p, \neg, \wedge$ and:

- $w \vDash \varphi > \psi$ iff $\psi \in F(w, \varphi)$.

A sentence selection model can be simulated in the revision operator semantics, by a *simple revision model* $\langle S, R^>, E, V \rangle$.

Theorem 4 *Any simple revision model $\mathfrak{M}^*$ is point-equivalent to a sentence selection model $\mathfrak{M}$. And conversely.*

Proof. Let $\mathfrak{M}^* = \langle S, R^>, E, V \rangle$ be a simple revision model. Set $W = S$ and $F(s, \varphi) = \bigcap_{s' \in R(s,\varphi)} \{\theta \in \mathcal{L} : s' E \theta\}$. Then $\langle W, F, V \rangle$ is a sentence selection model. By classicality of $\neg, \wedge$, we only need to check:

$$s \vDash_{\mathfrak{M}^*} \varphi > \psi \quad \text{iff} \quad \forall s' \in R^>(s, \varphi), s' E \psi \quad (>)$$
$$\text{iff} \quad \psi \in \bigcap_{s' \in R^>(s,\varphi)} \{\theta \in \mathcal{L} : s' E \theta\} \quad \text{(Set Theory)}$$
$$\text{iff} \quad \psi \in F(s, \varphi) \quad (F)$$

Revision Operator Semantics

Let $\mathfrak{M} = \langle W, F, V' \rangle$ be a sentence selection model. Define $\mathfrak{M}^* = \langle S, R^>, E, V \rangle$ as follows: Denote $P = W \times \mathcal{L}$ and instead of writing the elements $\langle w, \varphi \rangle$, we write them w_φ. For $s = \langle w, \varphi \rangle \in P$, the world part $s_0 = w$ and the sentence part $s_1 = \varphi$ are well defined. Set $S = W \cup P$. If $s \in W$, set $R^>(s, \varphi) = \{s_\varphi\}$; if $s \in P$, set $R(s, \varphi) = \{(s_0)_\varphi\}$. Thus, for $w \in W$, we obtain $R(w_\chi, \varphi) = R(w, \varphi) = \{w_\varphi\} \subseteq P$. Furthermore, for $s \in W$, define $V(s) := V'(s)$, and for $s \in P$, define $V(s) := V(s_0)$. Finally, for $s \in P$, $sE\psi$ iff $\psi \in F(s_0, s_1)$. For $s \in W$, the value of $sE\psi$ is unimportant and may be chosen as "true", since sE will never be considered for such $s \in W$, because we have $R(s, \varphi) \subseteq P$! Note that for any fixed $w \in W$, w and w_χ are indistinguishable in truth values: this is clear for p, $\neg$ and $\wedge$ by definition of V ($V(w_\chi) = V((w_\chi)_0) = V(w)$) and classicality of $\neg$ and $\wedge$. This extends to $>$:

$w_\chi \vDash_{\mathfrak{M}^*} \varphi > \psi$ iff $\forall s' \in R^>(w_\chi, \varphi), s'E\psi$ (>)

           iff $w_\varphi E\psi$ ($R^>(w_\chi, \varphi) = \{w_\varphi\}$)

           iff $\forall s' \in R^>(w, \varphi), s'E\psi$ ($R^>(w, \varphi) = \{w_\varphi\}$)

           iff $w \vDash_{\mathfrak{M}^*} \varphi > \psi$ (>)

For point-equivalence, consider the identity function $W \mapsto W$ and:

$w \vDash_{\mathfrak{M}} \varphi > \psi$ iff $\psi \in F(w, \varphi)$ (Def)

           iff $w_\varphi E\psi$ (E)

           iff $\forall s' \in R^>(w, \varphi), s'E\psi$ ($R^>$)

           iff $w \vDash_{\mathfrak{M}^*} \varphi > \psi$ (>)

$\square$

The proof shows that it does not matter whether we adopt the precise or the imprecise Ramsey Test, i.e., whether we consider simple revision models $\langle S, R^>, E, V \rangle$ or *precise* simple revision models, where each $R^>(s, \varphi)$ is a singleton. This is contrary to *set-selection models* $\langle W, f, V \rangle$. In these the conditional is defined as $w \vDash_{\mathfrak{M}} \varphi > \psi$ iff $f(w, [\varphi]) \subseteq [\psi]$, for $f : W \times \wp(W) \longrightarrow \wp(W)$. Here the difference between the imprecise and precise Ramsey Test, i.e., between $f(w, [\varphi])$ a set or a singleton, shows up in a logical law. Precise set-selection models validate conditional excluded middle ($\varphi > \psi) \vee (\varphi > \neg\psi$). In simple revision models, this difference does not arise, as long as E is not assumed to be $\vDash$. Because, even if $R^>(s, \varphi) = \{s'\}$, we may have neither $s'E\psi$ nor $s'E\neg\psi$.

Let us examine semantics, where $>$ and $\neg$ are non-classical.

Example 4 (intuitionism) A Kripke model $\langle W, R, V \rangle$ is *intuitionistic* if R is a preorder (reflexive, transitive), $\wedge, \Box$ are defined as in a Kripke model, $>$ is strict implication, $\neg\varphi \equiv \varphi > \bot$, $w \vDash \bot$ for no w and

- if wRv and $w \in V(p)$ then $v \in V(p)$. (R-persistency)

Theorem 5 *Over $\mathcal{L}_\Box$, any intuitionistic Kripke model is point-equivalent to an intuitionistic revision model $\langle S, R^>, R^\neg, L, V \rangle$, $R^>$ based on a preorder $R^\geq_\top$, $R^\neg = R^>$, V $R^\geq_\top$-persistent and $s'L\varphi$ iff $s' \vDash \bot$. And conversely.*

Proof. Omitted. □

In other words, the intuitionistic conditional is a strict implication and intuitionistic negation an explicit conditional negation.

Another example of non-classical $>$ is relevance logic without negation (Dunn & Restall, 2002; Urquhart, 1972).

Example 5 (relevance) $\langle I, \circ, V \rangle$ is a *semi-lattice model*, if $I \neq \emptyset$ is a set of information pieces with $0 \in I$ the empty piece of information, $\circ$ combines information pieces and V is a valuation. Additionally, $\langle I, \circ, 0 \rangle$ is a bounded (join) semi lattice, i.e., $\circ$ is total, associative, commutative, idempotent ($x \circ x = x$) and 0 is the identity ($0 \circ x = x$).[11] The atomic clause and $\wedge$ are classical, there is no negation and $>$ is defined by

- $i \vDash \varphi > \psi$ iff for all $j \in I$, if $j \vDash \varphi$ then $i \circ j \vDash \psi$.

φ is *valid*, denoted $\vDash \varphi$, iff $0 \vDash_\mathfrak{M} \varphi$ for all semi-lattice models $\mathfrak{M}$.

Theorem 6 *In $\mathcal{L}_{\backslash\{\neg\}}$, any semi-lattice model is point-equivalent to a revision model.*

Proof. Let $\mathfrak{M} = \langle I, \circ, V \rangle$ be a semi-lattice model. Define $\mathfrak{M}^* = \langle S, R^>, V \rangle$ with $S = I$, $R^>(s, \varphi) = \{s \circ t : \exists t \in S, t \vDash \varphi\}$ (E is $\vDash$). It suffices to check

$$\begin{aligned}
i \vDash_\mathfrak{M} \varphi > \psi \quad &\text{iff} \quad \forall j(j \vDash_\mathfrak{M} \varphi \Rightarrow i \circ j \vDash_\mathfrak{M} \psi) &&\text{(Def)}\\
&\text{iff} \quad \forall s' \forall j((j \vDash_\mathfrak{M} \varphi \,\&\, s' = i \circ j) \Rightarrow s' \vDash_\mathfrak{M} \psi)\\
&\text{iff} \quad \forall s' \forall j((j \vDash_{\mathfrak{M}^*} \varphi \,\&\, s' = i \circ j) \Rightarrow s' \vDash_{\mathfrak{M}^*} \psi) &&\text{(IH)}\\
&\text{iff} \quad \forall s' \in R^>(i, \varphi), s' E \psi &&(R^>, E)\\
&\text{iff} \quad i \vDash_{\mathfrak{M}^*} \varphi > \psi &&(>)
\end{aligned}$$

□

---

[11] Equivalently $\langle I, \leq, 0 \rangle$ is a partial order with the least upper bound property and smallest element 0, where $x \leq y$ iff $x \circ y = y$.

Revision Operator Semantics

Let us examine a slightly deviant version, where no operator is classical.

Example 6 (update semantics) An *update model* has the form $\langle \wp(W), [\cdot], v \rangle$, where $W \neq \emptyset$, $v \colon \text{Var} \longrightarrow \wp(W)$ is a valuation, $s \in \wp(W)$ are states and the update function $[.] \colon \wp(W) \times \mathcal{L} \longrightarrow \wp(W)$ is:

- $s[p] = s \cap v(p)$,
- $s[\varphi \wedge \psi] = s[\varphi][\psi]$,
- $s[\neg \varphi] = s \setminus s[\varphi]$,
- $s[\varphi > \psi] = s$ if $s[\varphi][\psi] = s[\varphi]$, else $= \emptyset$.

Satisfaction ('support') is defined as $s \vDash \varphi$ iff $s[\varphi] = s$.

Note that updating is neither commutative, nor monotone, nor idempotent. Contrary to usual semantics, it is not satisfaction but updating which is inductively defined. However, for the language under consideration, one can show that conjunction is classical after all. For this, one proves:

Observation 1 *Over the language $\mathcal{L}$:*

1. $s[\varphi] \subseteq s$. 2. $\emptyset[\varphi] = \emptyset$. 3. $s[\varphi][\psi] = s$ iff $s[\varphi] = s = s[\psi]$.

Proof. (1). Inductively over the complexity of the formula, considering all states simultaneously. Denote IH the induction hypothesis:

$s[p] = s \cap v(p) \subseteq s$.

$s[\neg \varphi] = s \setminus s[\varphi] \subseteq s$. (No IH needed).

$s[\varphi \wedge \psi] = s[\varphi][\psi] \subseteq s[\varphi] \subseteq s$ (using the IH twice, once for $s'[\psi]$ with $s' = s[\varphi]$ and once for $s[\varphi]$).

$s[\varphi > \psi] = s$ or $s[\varphi > \psi] = \emptyset \subseteq s$. (No IH needed).

(2). $\emptyset[\varphi] \subseteq \emptyset$ (by 1). Thus $\emptyset[\varphi] = \emptyset$.

(3). By (1): $s[\varphi][\psi] = s$ iff $s[\varphi] = s$ & $s[\varphi][\psi] = s$ iff $s[\varphi] = s = s[\psi]$. □

We can then reformulate the support clauses recursively:

Observation 2 *In the language $\mathcal{L}$, we have:*

1. $s \vDash p$ iff $s \subseteq v(p)$,
2. $s \vDash \neg \varphi$ iff $s[\varphi] = \emptyset$,
3. $s \vDash \varphi \wedge \psi$ iff $s \vDash \varphi$ and $s \vDash \psi$,
4. $s \vDash \varphi > \psi$ iff $s[\varphi] \vDash \psi$.

Proof. We use Observation 1.

(1). $s \vDash p$ iff $s[p] = s$ iff $s \cap v(p) = s$ iff $s \subseteq v(p)$.

(2). $s \vDash \neg \varphi$ iff $s \setminus s[\varphi] = s$ iff $s[\varphi] = \emptyset$ (Observation 1.1).

(3). $s \vDash \varphi \wedge \psi$ iff $s[\varphi \wedge \psi] = s$ $\hfill (\vDash)$
  iff $s[\varphi][\psi] = s$ $\hfill (s[\varphi \wedge \psi])$
  iff $s[\varphi] = s$ and $s[\psi] = s$ $\hfill$ (Observation 1.3)
  iff $s \vDash \varphi$ and $s \vDash \psi$ $\hfill (\vDash)$

(4). $s \vDash \varphi > \psi$ iff $s[\varphi > \psi] = s$ $\hfill (\vDash)$
  iff $s = \emptyset$ or $s[\varphi][\psi] = s[\varphi]$ $\hfill (s[\varphi > \psi])$
  iff $s[\varphi][\psi] = s[\varphi]$ $\hfill$ (Observation 1.2)
  iff $s[\varphi] \vDash \psi$ $\hfill (\vDash)$

$\square$

Thus update conjunction is classical after all. The update conditional is based on a precise Ramsey Test. And update negation is an explicitly definable conditional negation, since by Observation 1.2 and 2.4: $s[\varphi] = \emptyset$ iff $s[\varphi][\bot] = s[\varphi]$ iff $s[\varphi] \vDash \bot$ iff $s \vDash \varphi > \bot$.

Update semantics can be simulated in *update* revision models $\langle S, R^x, L, V \rangle$, where $S = \wp(W)$, $V(s) = \{p : s \subseteq v(p)\}$, $R^x(s, \varphi) = \{s[\varphi]\}$, and $s'L\varphi$ iff $s' = \emptyset$ (E and M are classical).

Theorem 7 *Each update model is point equivalent to an update revision model.*

Proof. We use Observation 2. Denote $R = R^x$ with $R(s, \varphi) = \{s[\varphi]\}$.
$s \vDash p$ iff $s \subseteq v(p)$ iff $p \in V(s)$. By Observation 2.1.
$s \vDash \neg \varphi$ iff $s[\varphi] = \emptyset$ iff $\forall s' \in R(s, \varphi), s'L\varphi$. By Observation 2.2.
$s \vDash \varphi \wedge \psi$ iff $s[\varphi][\psi] = s$ iff $\exists s' \in R(s, \varphi)\, \exists s'' \in R(s', \psi), s''Ms$.
$s \vDash \varphi > \psi$ iff $s[\varphi] \vDash \psi$ iff $\forall s' \in R(s, \varphi), s'E\psi$. By Observation 2.4. $\square$

Update semantics does it all with just one precise revision operator. Because of the above Observation, we could have treated $\neg$ as definable and $\wedge$ as classical. Thus for the comparison so far, we could have used classical conjunction all the way. However, in some dynamic semantics (with additional modal operators), conjunction seizes to be classical and it was to incorporate these that we formulated conjunction revision-theoretically.

To illustrate *intensional conjunction*, consider exact truth maker semantics (Fine, 2012). For this we extend M to a ternary relation $s's''Ms$.

Revision Operator Semantics

Example 7 (truthmaking) $\langle S, \sqsubseteq, |\cdot|\rangle$ is a *state-model* if $S \neq \emptyset$, $\langle S, \sqsubseteq\rangle$ is a join semilattice[12], and $|\cdot|: \text{Var} \longrightarrow \wp(S)^2$, with $|p|_0$ the verifiers and $|p|_1$ the falsifiers. *Verification* and *falsification* are defined as follows:

- $s \Vdash p$ iff $s \in |p|_0$,
- $s \dashv\vdash p$ iff $s \in |p|_1$,
- $s \Vdash \neg\varphi$ iff $s \dashv\vdash \varphi$,
- $s \dashv\vdash \neg\varphi$ iff $s \Vdash \varphi$,
- $s \Vdash \varphi \wedge \psi$ iff for some t, u, $t \Vdash \varphi$, $u \Vdash \psi$ and $t \sqcup u = s$,
- $s \dashv\vdash \varphi \wedge \psi$ iff $s \dashv\vdash \varphi$ or $s \dashv\vdash \psi$.

A *restricted falsifier* revision model for $\mathcal{L}_{\setminus\{>\}}$ is of the form $\langle S, R^\wedge, L, M, V\rangle$, without $R^>$ and E, and such that $R^\wedge(s, \varphi) = [\varphi]$, M is ternary and determines an idempotent commutative semigroup over S,[13] with $x \sqcup y = z$ iff $xyMz$, and L is falsifying, i.e., $sL\neg\varphi$ iff $s \models \varphi$, $sL(\varphi \wedge \psi)$ iff $sL\varphi$ or $sL\psi$.

Theorem 8 *For every state-model there is a point-equivalent restricted falsifier revision model for $\mathcal{L}_{\setminus\{>\}}$. And conversely.*

Proof. Let $\langle S, \sqsubseteq, |\cdot|\rangle$ be a state-model. Define the revision model $\langle S, R^\wedge, L, M, V\rangle$, **without** $R^>$ and E, and where $R^\neg$ is classical, as follows: $V(s) = \{p : s \in |p|_0\}$; $s's''Ms$ iff $s' \sqcup s'' = s$; $R^\wedge(s, \varphi) = [\varphi]$, as well as: sLp iff $s \in |p|_1$, $sL\neg\varphi$ iff $s \models \varphi$, and $sL(\varphi \wedge \psi)$ iff $sL\varphi$ or $sL\psi$. We show by induction: $s \Vdash \alpha$ iff $s \models \alpha$; and $s \dashv\vdash \alpha$ iff $sL\alpha$.

$s \Vdash p$	iff $s \in |p|_0$	(Def)		$s \dashv\vdash p$ iff $s \in |p|_1$		(Def)
	iff $p \in V(s)$	(V)		iff sLp		(L)
	iff $s \models p$	(p)				
$s \Vdash \neg\varphi$	iff $s \dashv\vdash \varphi$	(Def)		$s \dashv\vdash \neg\varphi$ iff $s \Vdash \varphi$		(Def)
	iff $sL\varphi$	(IH)		iff $s \models \varphi$		(IH)
	iff $\forall s' \in R^\neg(s, \varphi), s'L\varphi$	$(R^\neg)$		iff $sL\neg\varphi$		(L)
	iff $s \models \neg\varphi$	$(\neg)$				
$s \Vdash \varphi \wedge \psi$	iff $\exists t\, \exists u, t \models \varphi, u \models \psi, t \sqcup u = s$					(Def, IH)
	iff $\exists t \in R^\wedge(s, \varphi)\, \exists u \in R^\wedge(t, \psi), tuMs$					$(R^\wedge, M)$
	iff $s \models \varphi \wedge \psi$					$(\wedge)$

---

[12] I.e., $\sqsubseteq$ is a partial order with all least upper bounds $s \sqcup s'$.
[13] $\langle S, *\rangle$ is a semigroup if $*$ is a total, binary and associative operation over S.

$s \dashv\vert \varphi \wedge \psi$ iff $s \dashv\vert \varphi$ or $s \dashv\vert \psi$ (Def)

iff $sL\varphi$ or $sL\psi$ (IH)

iff $sL(\varphi \wedge \psi)$ (L)

Conversely, let $\langle S, R^{\wedge}, L, M, V \rangle$ be a restricted falsifier revision model. Define the state-model $\langle S, \sqsubseteq, |\cdot| \rangle$, with $s' \sqcup s'' = s$ iff $s's''Ms$. $s \sqsubseteq t$ iff $s \sqcup t = t$. $s \in |p|_1$ iff sLp; $s \in |p|_0$ iff $p \in V(s)$. Point-equivalence can then be shown as above. □

Denote $\mathcal{L}^-$ the language restricted to simple counterfactuals $\varphi > \psi$, i.e., where neither φ nor ψ contain $>$. Fine (2012) suggests analysing these in world states:

- $w \Vdash \varphi > \psi$ iff $\forall t \, \forall u \, ((t \Vdash \varphi \, \& \, t \to_w u) \Rightarrow u \, \Vert{>} \, \psi)$.

The transition $t \to_w u$ represents that imposing t on w has as possible outcome u. Inexact verification is defined as

- $u \, \Vert{>} \, \psi$ iff for some $u' \sqsubseteq u$, $u' \Vdash \psi$.

Thus, we can rephrase the counterfactual clause as

- $w \Vdash \varphi > \psi$ iff $\forall t \, \forall u \, ((t \Vdash \varphi \, \& \, t \to_w u) \Rightarrow \exists u' \sqsubseteq u \, (u' \Vdash \psi))$.

Theorem 9 *The first point-equivalence of Theorem 8 extends to $\mathcal{L}^-$.*

Proof. We extend the previous restricted falsifier revision model by adding $R^>$ and E (both restricted to $\mathcal{L}_{\setminus\{>\}}$) to obtain $\langle S, R^>, R^{\wedge}, E, L, M, V \rangle$: $R^>(s,\varphi) = \{u \in S : \exists t \in S, t \vDash \varphi, t \to_s u\}$. $uE\psi$ iff there is u' such that $u' \vDash \psi$ and $u'uMu$ (i.e., $u' \sqsubseteq u$).

$w \Vdash \varphi > \psi$ iff $\forall t \forall u \, ((t \Vdash \varphi \, \& \, t \to_w u) \Rightarrow \exists u' \sqsubseteq u(u' \Vdash \psi))$ (Def)

iff $\forall u \forall t ((t \vDash \varphi \, \& \, t \to_w u) \Rightarrow uE\psi)$ (IH E)

iff $\forall u \in R^>(w,\varphi), uE\psi$ (∗)

iff $w \vDash \varphi > \psi$ (>)

Let us show (∗). ($\Rightarrow$): Let $u \in R^>(w,\varphi)$. Then there is $t \vDash \varphi$ such that $t \to_w u$ (Def $R^>$). ($\Leftarrow$): Let t, u such that $t \vDash \varphi$ and $t \to_w u$. Then, by definition of $R^>$, we also have $u \in R(s, \varphi)$. □

In brief, in the so-constructed revision model, E represents the modal "inexact verification" and $R^>(w,\varphi)$ is a derived revision operator, based on the transition relation $\to_w$.

5 Conclusion

Revision operator semantics is a proposal to unify known semantics and their logics, such as classical logic, intuitionistic logic, non-monotonic conditional logics, relevance logics, and logics arising from update or truthmaker semantics. The revision conditional generalises the Ramsey Test to a more general hypothetical test. The revision negation generalises modal and conditional negation. The revision conjunction generalises update conjunction and intensional conjunction. Classical connectives can always be obtained as well. By proving point-equivalences, I have shown how different known semantics relate to the revision operator semantics. An obvious corollary, are the analogue relations between the corresponding logics.

References

Berto, F. (2015). A Modality Called 'Negation'. *Mind*, *124*(495), 761–793.

Chellas, B. F. (1975). Basic conditional logic. *Journal of Philosophical Logic*, *4*(2), 133–153.

Dunn, M., & Restall, G. (2002). Relevance logic. In D. Gabbay & F. Guenthner (Eds.), *Handbook of Philosophical Logic*. Kluwer Academic Publishers.

Fine, K. (2012). Counterfactuals without possible worlds. *The Journal of Philosophy*, *109*, 221–246.

Gillies, A. S. (2004). Epistemic conditionals and conditional epistemics. *Noûs*, *38*(4), 585–616.

Leitgeb, H. (2019). Hype: A system of hyperintensional logic (with an application to semantic paradoxes). *Journal of Philosophical Logic*, *48*(2), 305–405.

Lewis, D. (1973). *Counterfactuals*. Cambridge MA: Harvard University Press.

MacColl, H. (1908). 'If' and 'Imply'. *Mind*, *XVII*(1), 151–152, 453–455.

Mares, E. D. (2012). Relevance and conjunction. *Journal of Logic and Computation*, *22*(1), 7–21.

Ramsey, F. P. (1929, 1990). General propositions and causality. In H. A. Mellor (Ed.), *Philosophical Papers*. Cambridge: Cambridge University Press.

Ripley, D. (2012). *Weak Negations and Neighborhood Semantics*. Retrieved from https://philpapers.org/rec/RIPWNA

Routley, R., Meyer, R., Plumwood, V., & Brady, R. (1983). *Relevant Logics and its Rivals, Volume I*. Atascardero, CA: Ridgeview.

Stalnaker, R. C. (1968). A theory of conditionals. *American Philosophical Quarterly*, 98–112.

Urquhart, A. (1972). Semantics for relevant logics. *Journal of Symbolic Logic*, 37(1), 159–169.

Veltman, F. (1996). Defaults in update semantics. *Journal of Philosophical Logic*, 25(3), 221–261.

Eric Raidl
University of Tuebingen
Excellence Cluster "Machine Learning: New Perspectives for Science"
Germany
E-mail: `eric.raidl@uni-tuebingen.de`

Strong Normalization in Core Type Theory

DAVID RIPLEY

Abstract: This paper presents a novel typed term calculus and reduction relation for it, and proves that the reduction relation is strongly normalizing—that there are no infinite reduction sequences. The calculus bears a close relation to the $\rightarrow, \neg$ fragment of core logic, and so is called 'core type theory'.

This paper presents a novel typed term calculus and reduction relation for it, and proves that the reduction relation is strongly normalizing—that there are no infinite reduction sequences. The calculus is similar to the simply-typed lambda calculus with an empty type, but with a twist. The simply-typed lambda calculus with an empty type bears a close relation to the $\rightarrow, \bot$ fragment of intuitionistic logic (Howard, 1980; Scherer, 2017; Sørensen & Urzyczyn, 2006); the calculus to be presented here bears a similar relation to the $\rightarrow, \neg$ fragment of a logic known as *core logic*. Because of this connection, I'll call the calculus *core type theory*.

Core logic (fka 'intuitionistic relevant logic') has been developed and studied primarily by Neil Tennant, in a series of papers spanning recent decades (see, e.g., Petrolo & Pistone, 2019; Tennant, 1979, 2002, 2017). One interesting feature of core logic is that its valid arguments are not closed under the rule of cut. That is, there are core valid arguments $[\Gamma \succ A]$ and $[A, \Delta \succ B]$ such that $[\Gamma, \Delta \succ B]$ is not core valid. However, in any such case, there is some subsequent of $[\Gamma, \Delta \succ B]$ that is core valid. Tennant calls this latter property *epistemic gain*, and maintains that usual motivations for closure under cut are at least as well served by epistemic gain.

This paper does not directly take a stand on that issue. However, at least one going motivation for cut is its connection to computational properties of a system, and in particular to reductions in related term calculi. For example, this is the connection Girard has in mind in saying that a "sequent calculus without cut-elimination is like a car without [an] engine" (Girard, 1995). So in evaluating Tennant's defense of core logic via epistemic gain, it would be

$$\to\text{I}^n\colon\ \frac{[\varphi]^n \vdots \psi}{\varphi\to\psi} \qquad \to\text{I!}^n\colon\ \frac{[\varphi]^n \vdots \odot}{\varphi\to\psi} \qquad \to\text{E}\colon\ \frac{\varphi\to\psi \quad \varphi}{\theta}\ \frac{[\psi]^n \vdots \theta}{}$$

$$\neg\text{I}^n\colon\ \frac{[\varphi]^n \vdots \odot}{\neg\varphi} \qquad \neg\text{E}\colon\ \frac{\neg\varphi \quad \varphi}{\odot}$$

In the rules →I! and ¬I, discharge may not be vacuous. In the rule →I, it may be vacuous.

Figure 1: Core natural deduction for intutionistic logic

useful to know what the situation is in related term calculi. This paper aims to shed light on that issue, by exploring some of the properties of core type theory.

1 Core logic

Although core logic is typically presented as a full first-order logic, this paper only studies its propositional →, ¬ fragment. As such, I won't present core logic in full here, just the needed fragment. Similarly, discussions of intuitionistic logic, which is closely related, will also focus on the propositional →, ¬ fragment. I will work with the natural deduction system presented in Fig. 1.

Since we're working in a fragment of the language without any falsum connective, there are questions about how to handle negation. We do so via ☺; this is not a formula or connective in its own right (and so cannot be used to form complex formulas), but rather a structural marker that interacts with the connective rules in the specified way. Not all nodes in a proof, then, must be occupied by formulas. Some can be occupied by the structural marker ☺ instead. However, while ☺ may appear in the course of a proof, it may not be assumed; only formulas are eligible for assumption.

For reasons that will become apparent later, I will use the term 'hat' to pick out candidate occupants of proof nodes. That is, every formula is a hat, and ☹ is a hat, and nothing else is a hat. I use Greek letters φ, ψ for formulas, and $\mathfrak{C}, \mathfrak{D}$ for hats. It is useful to consider hats as ordered, with $\mathfrak{C} \leq \mathfrak{D}$ iff either $\mathfrak{C}$ is ☹ or $\mathfrak{C} = \mathfrak{D}$.

With that qualification understood, but otherwise read straightforwardly, this natural deduction system determines not core logic but intuitionistic logic.

Definition 1 *For a set Γ of formulas and any $\mathfrak{C}$, we say $\Gamma \vdash \mathfrak{C}$ iff there is a proof of $\mathfrak{C}$ in the system of Fig. 1 whose open assumptions are all in Γ.*

Theorem 1 *The proof system in Fig. 1 determines intuitionistic logic. That is, $\Gamma \vdash \varphi$ iff $\Gamma \vdash_{Int} \varphi$; and $\Gamma \vdash ☹$ iff $\Gamma \vdash_{Int}$.*

Proof. It should be clear that the system does not prove anything that is intuitionistically unprovable. Note for later in the proof: this means that there is no proof of ☹ without open assumptions, since such a proof would require proofs with no open assumptions of both $\neg\varphi$ and φ for some φ, but intuitionistic logic is consistent. It remains to show that the system proves enough.

$\to$I and $\to$E together are usual rules for intuitionistic implication. $\neg$I and $\neg$E together are *almost* usual rules for intuitionistic negation; the difference is the prohibition on vacuous discharge in $\neg$I. So it is enough to show that an unrestricted version of $\neg$I is derivable in this system.

Suppose, then, that we have some proof Π of ☹, with an aim to proving $\neg\varphi$ regardless of whether φ is among the open assumptions of Π. As we've observed, Π must have some open assumption or other; say ψ is one. Let Π' be the proof of $\psi \to \neg\varphi$ arrived at by extending Π with a step of $\to$I!, discharging the occurrences of ψ open in Π. Then we can proceed as follows:

$$\to E^n: \quad \frac{\overset{\Pi'}{\psi \to \neg\varphi} \quad \psi \quad [\neg\varphi]^n}{\neg\varphi}$$

The overall proof arrived at has exactly the same open assumptions as Π, and is a proof of $\neg\varphi$, regardless of whether φ is among the open assumptions of Π. □

Why give such a strange proof system for good old intuitionistic logic? Because this system makes it particularly simple to reach core logic as well. To do this, we impose an additional restriction, which I will call the *core restriction*.

Definition 2 *A proof meets the* core restriction *iff in each application of $\to$E and $\neg$E in the proof, the major premise of the application is itself assumed, rather than following from another rule.*[1] *A proof is a* core proof *iff it meets the core restriction.*

For a set Γ of formulas and a formula φ, we say $\Gamma \vdash^- \varphi$ iff there is a core proof of φ whose open assumptions are all in Γ.

Not everything provable in the total system has a core proof. (Note that the crucial derivation invoked in the proof of Theorem 1 is not core.) Core logic is the logic determined by core proofs.

Example 1 $\neg\varphi, \varphi \vdash \psi$, but $\neg\varphi, \varphi \not\vdash^- \psi$. Here is a proof establishing the first claim:

$$\neg\text{E:} \quad \frac{\neg\varphi \quad [\varphi]^1}{\odot}$$
$$\to\text{I!}^1: \quad \frac{}{\varphi \to \psi}$$
$$\to\text{E}^2: \quad \frac{\varphi \to \psi \quad \varphi \quad [\psi]^2}{\psi}$$

Note that the proof is not core, since the major premise for $\to$E is not an assumption.

To see that there is no core proof of the same, note first that there are no implications in the conclusion of this proof. Since no implication introduced in the course of a core proof can ever be eliminated, this must mean that the rules $\to$I and $\to$I! are not involved in any core proof with this conclusion. Similar reasoning shows that $\neg$I cannot be involved either. But if $\to$I! and $\neg$I are neither of them involved, then $\neg$E cannot be involved, since it would bring a $\odot$ into the proof, and $\to$I! and $\neg$I are the only rules that can continue on from a $\odot$. So any purported core proof for this argument must consist entirely of $\to$E, and as such must leave some assumption open of the form $\rho \to \theta$. But neither premise of this argument has such a form; so there is no such proof.

It should be noted that this proof system for core logic is not the exact system presented and studied by Tennant (see, e.g., Tennant, 2017). In Tennant's system, any occurrence of any assumption that can be discharged by a rule application *must* be discharged by that rule application. I have not imposed this restriction. This means that core proofs as Tennant defines them are not closed under substitution, since a proof that violates this restriction

---

[1]In $\to$E, the major premise is the displayed $\varphi \to \psi$; and in $\neg$E, the major premise is the displayed $\neg\varphi$.

can be a substitution instance of a proof that meets it. For this reason, I have here dropped the restriction. However, this does not affect which arguments have core proofs; it only affects how many proofs they have. This is because while *proofs* are not closed under substitution when Tennant's restriction is imposed, *provability* still is. So closing proofs themselves under substitution, which is the effect of relaxing this restriction, does not affect provability.[2,3]

Finally, it's worth noting a key relation between intuitionistic logic and core logic, and between these proof systems for the logics. For more information, see (Tennant, 2002, 2015).

Theorem 2 *Any proof of $\mathfrak{C}$ with open assumptions Γ can be normalized into a core proof of $\mathfrak{D}$ with open assumptions Δ, for some $\mathfrak{D}, \Delta$ with $\mathfrak{D} \leq \mathfrak{C}$ and $\Delta \subseteq \Gamma$.*

It follows from this that whenever $\Gamma \vdash \varphi$, either $\Gamma \vdash^- \varphi$ or $\Gamma \vdash^- \odot$, and moreover that $\Gamma \vdash \odot$ iff $\Gamma \vdash^- \odot$. So core logic matches intuitionistic logic on the consequences of consistent sets, and it matches intuitionistic logic in its understanding of which sets are consistent in the first place. The only differences between intuitionistic validity and core validity are in what follows from inconsistent sets. Intuitionistically, everything follows; this is not so in core logic.

2 Core type theory

2.1 Types and hats

The propositions of core logic will serve as the types of core type theory. There will also be need of a slight generalisation of types, what I will call *hats*. A *hat* is either a type or else $\odot$.[4] I will continue to use lowercase Greek letters to indicate propositions (and so types); for hats I use $\mathfrak{C}, \mathfrak{D}$. Hats are considered to be ordered: $\mathfrak{C} \leq \mathfrak{D}$ iff either $\mathfrak{C} = \mathfrak{D}$ or $\mathfrak{C} = \odot$.

---

[2] Thanks to Neil Tennant (pc) for discussion on this point.

[3] A preview and a conjecture: this means that the corresponding restriction in the term system to follow would not affect which types are inhabited, but only how many inhabitants they have. I would conjecture that the term calculus that corresponds to Tennant's actual system, with this extra discharge restriction imposed, is the fragment of the system presented in Section 2 involving only one variable of each type, rather than denumerably many.

[4] Just as in the proof system of Fig. 1, note that $\varphi \to \odot$, $\neg \odot$, and so on are not well-formed. The only way $\odot$ can appear in a hat is alone. $\to$ and $\neg$ operate on *types*, and $\odot$ is not a type.

David Ripley

Each term in core type theory wears a (unique) hat.[5] If the term's hat is a type, the term is *typed*; if the hat is ☺, the term is *exceptional*. Following usual interpretations of the Curry-Howard correspondence (e.g., Sørensen & Urzyczyn, 2006), terms can be understood as programs. When the term's hat is a type φ, this indicates that if the program is successfully run its output will be data of that type. When the term's hat is ☺, this indicates that if the program is run it will not be successful.

This requires interpretations of the complex types $\varphi \to \psi$ and $\neg \varphi$. Implications are interpreted as function types. A function of type $\varphi \to \psi$ awaits an input of type φ, and if run successfully with such an input, it produces an output of type ψ. Negations are interpreted as failures triggered by the presence of some other data. A term of type $\neg \varphi$ awaits an input of type φ, and if it receives such an input it fails to run successfully.

For now, these explanations are just intuitive indications. They will be made precise in Sections 2.2 and 2.4.

2.2 Terms

Terms (with their hats) are defined inductively as follows. Here and throughout, where I omit a hat on a term, either it can be inferred or I am speaking in generality. For a term M, $\text{FV}(M)$ is the set of free variables in M; this is defined as usual. These term-forming clauses are directly determined by copying the proof rules of Fig. 1, with variables standing for assumptions.

Definition 3 (Terms)

- *There's a countable infinity of variables $x^\varphi, y^\varphi, \ldots$ of each type φ.*

- *Given $M^{\varphi \to \psi}$ and N^φ, we have $(MN)^\psi$.*

- *Given $M^{\neg \varphi}$ and N^φ, we have $(MN)^\odot$*

- *Given x^φ and M^ψ, we have $(\lambda x.M)^{\varphi \to \psi}$, in which any and all free occurrences of x in M are bound.*

- *Given x^φ and $M^\odot$ with x free, we have $(\lambda x.M)^{\varphi \to \psi}$ and $(\lambda x.M)^{\neg \varphi}$, in which all free occurrences of x in M are bound.*

Some comments on this definition are in order. First, note that there are no variables with ☺ as a hat. Variables must be *typed*, not merely hatted. Second,

---

[5]This is thus a 'Church-style' and not a 'Curry-style' calculus, in the language of Sørensen and Urzyczyn (2006, p. 63).

Strong Normalization in Core Type Theory

note that there is never any restriction on binding multiple occurrences of the same variable; this is always fine. Third, note that *vacuous* binding is sometimes allowed and sometimes not. Vacuously binding into a *typed* term is fine, but vacuously binding into an exceptional term is not possible. Fourth, note that $(\lambda x.M)^{\varphi \to \psi}$ and $(\lambda x.M)^{\neg \varphi}$ are distinct terms (when they are well-formed), even though they differ only in their hats.

In everything that follows, terms are everywhere identified up to α-conversion (relettering of bound variables). These term-formation rules can be seen as analogous to the natural deduction rules of Fig. 1, with the difference that I've here chosen the term-formation rule analogous to a more usual formulation of $\to$E.

2.3 Substitution

Substitution (of a term for a variable) is defined as usual. It is well-defined iff the term's hat matches the type of the variable it is being substituted for. Since there are no exceptional variables, exceptional terms cannot be substituted for anything. I write $M[x \mapsto N]$ to indicate the term that results from substituting N for each free occurrence of x in M.

2.4 Reduction

2.4.1 Redexes and reducts

A *redex* is a term of the form $(\lambda x.M)N$, a lambda abstract applied to an argument. In any redex, the hat on N matches the hat on x, and so $M[x \mapsto N]$ is defined; this is the *reduct* of the redex.

There are two things to notice here:

- The free variables in a redex are a superset of the free variables in its reduct. The superset is possibly proper, owing to the possibility of vacuous binding: if x is not free in M, then N's free variables may not occur at all in $M[x \mapsto N]$.

- The hat on a redex is $\geq$ the hat on its reduct. The order is possibly proper: $((\lambda x.M^\ominus)^{\varphi \to \psi} N)^\psi$ has as its reduct $(M^\ominus[N \mapsto x])^\ominus$.

The second remark here reveals that all is not as usual. Already we can see that core type theory fails to exhibit one usual property of type term calculi, variously called 'preservation' or 'subject reduction' (see, e.g., Barthe & Melliès, 1996; Harper, 2016). Typed redexes can have exceptional reducts.

David Ripley

However, the second remark also shows that the situation is not entirely unconstrained, at least for redexes and their reducts. We can never move from one type to another, or from ☺ to a type.

Definition 4 *In a redex* $(\lambda x.M)N$, *if x does not occur free in M, the redex is a* vacuous *redex. If M is $M^{\ominus}$, the redex is an* explosive *redex.*

Note that no redex can be both vacuous and explosive, although some are neither.

2.4.2 One-step reduction

This subsubsection defines the relation $\triangleright_{1\beta}$ of *one-step* reduction between terms. To begin: if M is a redex and M' its reduct, then $M \triangleright_{1\beta} M'$.

In more usual calculi, such as the simply-typed lambda calculus, reduction is a *compatible* relation; if $M \triangleright_{1\beta} M'$, then $O(M) \triangleright_{1\beta} O(M')$. In such calculi, then, we can reduce a complex term containing a redex simply by reducing the redex itself, leaving the rest alone. Indeed, this is required for a calculus to be a term rewriting system, in the sense of Baader and Nipkow (1998); Terese (2003).

In core type theory, however, this is not possible.[6] There are two reasons for this. First, we can have $M \triangleright_{1\beta} M'^{\ominus}$ where $x \notin \text{FV}(M')$ and either $x \in \text{FV}(M)$ or M is M^{φ}. In either case, $\lambda x.M$ is well-formed but $\lambda x.M'$ is not. In $\lambda x.M$ either the binding is not vacuous or M is not exceptional, so there is no problem. But $\lambda x.M'$ attempts to vacuously bind into an exceptional term, which is not possible. So we cannot reduce terms underneath a lambda without exercising some care.

Second, in terms of the form MN, both M and N must be typed. If $M \triangleright_{1\beta} M'^{\ominus}$, then $M'N$ is not well-formed; and if $N \triangleright_{1\beta} N'^{\ominus}$, then MN' is not well-formed. So we cannot reduce either term in an application without exercising some care. As a result of these situations, the needed definition of one-step reduction is less straightforward than usual, and is given by recursion on the structure of the term being reduced.

Definition 5 (One-step reduction)

- *If M is a redex and M' its reduct, then $M \triangleright_{1\beta} M'$*

---

[6] And so core type theory, despite involving terms with familiar-seeming structure, is not a term rewriting system. Despite this, I have borrowed some language, such as 'redex', from the theory of term rewriting systems, since this language applies just as well to core type theory. Core type theory is still an abstract reduction system in the sense of Baader and Nipkow (1998); Terese (2003).

Strong Normalization in Core Type Theory

- *Reducing M in MN:*
  - If $M^\varphi \rhd_{1\beta} M'^\varphi$, then $MN \rhd_{1\beta} M'N$
  - If $M \rhd_{1\beta} M'^\ominus$, then $MN \rhd_{1\beta} M'$
- *Reducing N in MN:*
  - If $N^\varphi \rhd_{1\beta} N'^\varphi$, then $MN \rhd_{1\beta} MN'$
  - If $N \rhd_{1\beta} N'^\ominus$, then $MN \rhd_{1\beta} N'^\ominus$
- *Reducing M in $\lambda x.M$:*
  - If $M^\varphi \rhd_{1\beta} M'^\varphi$, then $\lambda x.M \rhd_{1\beta} \lambda x.M'$
  - If $M \rhd_{1\beta} M'^\ominus$,

    if $x \in \mathrm{FV}(M')$, then $\lambda x.M \rhd_{1\beta} \lambda x.M'$ (preserving hat[7])

    if $x \notin \mathrm{FV}(M')$, then $\lambda x.M \rhd_{1\beta} M'$

The strategy of this definition is straightforward: if it is possible to reduce a subterm in place without changing the context, then that's what's done. So when $M \rhd_{1\beta} M'$, if $O(M)$ and $O(M')$ are both well-formed, then indeed $O(M) \rhd_{1\beta} O(M')$. That much of compatibility stands. When this is not possible, because the result would not be well-formed, then the subterm being reduced is retained, and its immediate context discarded. There are no choices in this process, other than the initial choice of a redex to reduce. So given a term with a redex in it, there is a unique result of reducing the term at that redex.

It can be verified by inspection of this definition that the two preservation properties indicated above for redexes and their reducts extend to $\rhd_{1\beta}$. That is, where $M^{\mathfrak{C}} \rhd_{1\beta} N^{\mathfrak{D}}$, then $\mathrm{FV}(M) \supseteq \mathrm{FV}(N)$ and $\mathfrak{C} \geq \mathfrak{D}$. (That these preservation properties hold is required for Definition 5 to work, but it is straightforward to see that they do.)

Some examples will help give the flavour of this definition.

Example 2 $\left((\lambda y^\varphi.(x^{\neg\varphi}y^\varphi)^\ominus)^{\varphi\to\theta} z^\varphi\right)^\theta$ is a redex, and it reduces in one step to $(x^{\neg\varphi}z^\varphi)^\ominus$.

---

[7]If $x^\varphi \in \mathrm{FV}(M'^\ominus)$, then $(\lambda x^\varphi.M'^\ominus)^{\varphi\to\psi}$ and $(\lambda x^\varphi.M'^\ominus)^{\neg\varphi}$ are both well-formed, and differ only in their hat. This clause should be understood as saying $(\lambda x.M)^{\varphi\to\psi} \rhd_{1\beta} (\lambda x.M')^{\varphi\to\psi}$ and $(\lambda x.M)^{\neg\varphi} \rhd_{1\beta} (\lambda x.M')^{\neg\varphi}$. We do *not* have $(\lambda x.M)^{\varphi\to\psi} \rhd_{1\beta} (\lambda x.M')^{\neg\varphi}$ or $(\lambda x.M)^{\neg\varphi} \rhd_{1\beta} (\lambda x.M')^{\varphi\to\psi}$; changing hats here is disallowed.

Example 3 Let M be the redex from Example 2, and let M' be its reduct. Then $(\lambda w^\rho.M^\theta)^{\rho\to\theta} \triangleright_{1\beta} M'^\ominus$.

Note that while $(\lambda w^\rho.M^\theta)^{\rho\to\theta}$ is well-formed, despite the fact that $w \notin \mathrm{FV}(M)$, the same would not be true of $\lambda w.M'^\ominus$. Accordingly, the λw is dropped entirely in this step of reduction, and we reach M' by itself.

Example 4 With the same M and M', we have $\bigl(\lambda z^\varphi.(\lambda w^\rho.M^\theta)^{\rho\to\theta}\bigr)^{\varphi\to\rho\to\theta} \triangleright_{1\beta} (\lambda z^\varphi.M'^\ominus)^{\varphi\to\rho\to\theta}$. While the inner vacuous lambda has dropped out, as we saw in Example 3, the outer λz is not vacuous, and so it can bind into the exceptional M'. Accordingly, it is retained.[8]

Example 5 With the same M and M', we have $\bigl((\lambda w^\rho.M^\theta)^{\rho\to\theta}v^\rho\bigr)^\theta \triangleright_{1\beta} M'^\ominus$. As in Example 3, the vacuous binding must vanish. Here, the application too must vanish: $M'v$ would not be well-formed, so M' alone is retained.

Example 6 With the same M, we have $\bigl((\lambda w^\rho.M^\theta)^{\rho\to\theta}v^\rho\bigr)^\theta \triangleright_{1\beta} M^\theta$. This starts from the same term as Example 5, but reduces it at the outer (vacuous) redex, rather than the inner one. Since w does not occur free in M, this means that $M[w \mapsto v]$ is just M itself.

These examples start to reveal some of the complexity of one-step reduction in core type theory. They also show a bit about how the preservation properties operate in practice.

With one-step reduction in hand, we can give a standard definition of normal form:

Definition 6 *A term M is in* normal form *iff there is no M' with $M \triangleright_{1\beta} M'$; or equivalently, iff it does not contain a redex.*

2.4.3 Reduction

Finally, *reduction* $\triangleright_\beta$ is the reflexive transitive closure of $\triangleright_{1\beta}$. Again, it can be verified that the two preservation properties apply to $\triangleright_\beta$ as well: where $M^{\mathfrak{C}} \triangleright_\beta N^{\mathfrak{D}}$, then $\mathrm{FV}(M) \supseteq \mathrm{FV}(N)$ and $\mathfrak{C} \geq \mathfrak{D}$. Both of these properties will be appealed to repeatedly in what follows. I will also talk of 'reduction paths' in the usual way.

---

[8] Although $(\lambda z^\varphi.M'^\ominus)^{\neg\varphi}$ is well-formed, it cannot be reached by reduction from this term, as discussed in Footnote 7.

Strong Normalization in Core Type Theory

Example 7 With the same M and M' from Examples 5 and 6, we can see two reduction paths from $\left((\lambda w^\rho.M^\theta)^{\rho\to\theta}v^\rho\right)^\theta$ to $M'^\odot$. The first path is one step long, and is the reduction in Example 5. The second path is two steps long: its first step is the reduction in Example 6, and its second is the reduction in Example 2.

2.4.4 Reduction and substitution

There's enough here to prove a key lemma about the interaction between reduction and substitution that will be important later.

Lemma 1 (q.v. Hindley & Seldin, 2008, Lemma A1.12, p. 280) *If x and y are distinct variables and y does not occur free in N, then $Q[y \mapsto P][x \mapsto N]$ is $Q[x \mapsto N][y \mapsto P[x \mapsto N]]$.*

Proof. The only things that could have gone wrong are ruled out by assumption. □

Lemma 2 *If $M \triangleright_\beta M'$ and $N \triangleright_\beta N'$, and $M[x \mapsto N]$ and $M'[x \mapsto N']$ are both defined, then $M[x \mapsto N] \triangleright_\beta M'[x \mapsto N']$.*

Proof. Since terms are identified up to α-conversion, we are free to assume that that no variable occurring bound in M is free in N, and that x is not bound in M.

If $M[x \mapsto N]$ and $M'[x \mapsto N']$ are both defined, then x^φ, N^φ, and N'^φ, for some type φ.[9] It follows that $O[x \mapsto N]$ and $O[x \mapsto N']$ are defined for any term O; the latter will be appealed to below.

First we worry about the reduction path from N to N'. For any step $N_i \triangleright_{1\beta} N_j$ in this path, we must have neither N_i nor N_j exceptional, since N' is not exceptional. It follows that $M[x \mapsto N_i]$ and $M[x \mapsto N_j]$ are both defined. Because of this, $M[x \mapsto N_i] \triangleright_{1\beta} M[x \mapsto N_j]$, since one-step reduction happens without changing its context whenever this is possible, and we have seen it is possible in this case. So $M[x \mapsto N] \triangleright_\beta M[x \mapsto N']$.

Now we worry about the reduction path from M to M'. For any step $M_i \triangleright_{1\beta} M_j$ in this path, $M_i[x \mapsto N']$ and $M_j[x \mapsto N']$ are both defined, as mentioned above. This step must occur at a redex $(\lambda y.Q)P$ in M_i with reduct $Q[y \mapsto P]$ in M_j—call this the *key* redex. We work by induction on the

---

[9]Usual statements of this lemma for typed lambda calculi, such as (Hindley & Seldin, 2008, Lemma A1.15, p. 281), do not need to assume that $M'[x \mapsto N']$ is defined, since this follows in those settings from the other assumptions of the lemma. Here, though, it does not follow. Fortunately, everywhere we need to apply this lemma the extra assumption will be justified.

formation of M_i around the key redex, to show $M_i[x \mapsto N'] \triangleright_{1\beta} M_j[x \mapsto N']$.

- If M_i just is the key redex $(\lambda y.Q)P$, then $M_i[x \mapsto N']$ is $(\lambda y.(Q[x \mapsto N']))(P[x \mapsto N'])$. (It's important here that y is not x, but as y is bound in M we have this.) Thus, $M_i[x \mapsto N'] \triangleright_{1\beta} (Q[x \mapsto N'])[y \mapsto P[x \mapsto N']]$; call this latter $M^\star$.

  In this case, M_j is just the reduct $Q[y \mapsto P]$, and so $M_j[x \mapsto N']$ is $Q[y \mapsto P][x \mapsto N']$. By Lemma 1, this is $M^\star$, since we have what that Lemma needs about bound variables.

- If M_i is OR, with the key redex in O, then $O \triangleright_{1\beta} O'$ at the key redex. So M_j is either $O'R$ if O' is typed or else O' if O' is exceptional. Either way, what we need follows from the induction hypothesis and Definition 5.

- If M_i is OR, with the key redex in R, the reasoning is just the same as the previous case.

- If M_i is $\lambda z.O$, with the key redex in O, then $O \triangleright_{1\beta} O'$ at the key redex. So M_j is either O' if O' is exceptional and does not contain z free, or else $\lambda z.O'$ otherwise. Either way, what we need follows from the induction hypothesis and Definition 5, so long as $O'[x \mapsto N']$ is exceptional and does not contain z free iff the same is true of O'. For this, it suffices that z is not x and does not occur free in N'; since z is bound in M_i and so in M, this is taken care of. $\square$

3 Strong normalisation

In core type theory, every reduction path reaches a normal form; reduction in core type theory is *strongly normalising*. This section proves this claim. I will proceed closely following the strategy of Hindley and Seldin (2008, Appendix A3), who themselves follow the work of Tait (1967).[10] Key to the proof is the interplay between two properties of terms: *strong normalizability* and *strong computability*.

Definition 7 *A term is* strongly normalizing *(SN) iff every reduction path beginning from the term is finite.*

---

[10] See also (Troelstra & Schwichtenberg, 2000, §6.8, 6.12.2) and (Sørensen & Urzyczyn, 2006, §5.3.2–5.3.6), both also cited by Hindley and Seldin (2008) in connection with this proof.

Strong Normalization in Core Type Theory

Definition 8 *A term is* strongly computable *(SC) as follows (by induction on hat):*

- If M is a term of atomic type or an exceptional term, then M is SC iff it is SN.

- $M^{\varphi \to \psi}$ is SC iff for all SC terms N^{φ}, the term $(MN)^{\psi}$ is SC.

- $M^{\neg \varphi}$ is SC iff for all SC terms N^{φ}, the term $(MN)^{\ominus}$ is SC.

It will emerge over the course of the proof that every term is both SN and SC. The proof begins with a series of small lemmas, before moving to three larger lemmas. The eventual theorem—that every term is SN, and so that $\triangleright_\beta$ is strongly normalizing—follows quickly from two of these larger lemmas.

3.1 Smaller lemmas

First, two immediate consequences of Definition 8.

Lemma 3 *If M and N are both SC, and MN is well-formed, then MN is SC.*

Proof. Immediate from Definition 8. □

In the next Lemma and a few places to come, I'll make use of the fact that every type has the form $\mathfrak{C}_1 \to \ldots \to \mathfrak{C}_n \to \theta$, for some $n \geq 0$ and θ either atomic or $\neg \sigma$ for some type σ.

Lemma 4 $M^{\mathfrak{C}}$ *is SC iff:*

- *where $\mathfrak{C}$ is $\mathfrak{C}_1 \to \ldots \to \mathfrak{C}_n \to \theta$ and θ is atomic: for all SC $N_1^{\mathfrak{C}_1}, \ldots, N_n^{\mathfrak{C}_n}$, the term $(MN_1 \ldots N_n)^{\theta}$ is SC.*

- *where $\mathfrak{C}$ is $\mathfrak{C}_1 \to \ldots \to \mathfrak{C}_n \to \neg \sigma$: for all SC $N_1^{\mathfrak{C}_1}, \ldots, N_n^{\mathfrak{C}_n}, O^{\sigma}$, the term $(MN_1 \ldots N_n O)^{\ominus}$ is SC.*

Proof. Definition 8. □

Now, two Lemmas that allow us to infer SN for some terms from SN for others:

Lemma 5 *If M is SN, and N is a subterm of M, then N is SN as well.*

Proof. By noting that any infinite reduction sequence for N would give rise to an infinite reduction sequence for M. □

Lemma 6 *If $M[x \mapsto N]$ is SN, then M is SN as well.*

Proof. By noting that any infinite reduction sequence for M would give rise to an infinite reduction sequence for $M[x \mapsto N]$. □

3.2 Bigger lemmas and the result

That's enough to step to the three bigger Lemmas.

Lemma 7 (q.v. Hindley & Seldin, 2008, Lemma A3.10, p. 295) *Let $\mathfrak{C}$ be any hat. Then:*

1. *if $M^{\mathfrak{C}}$ is of the form $(aX_1 \ldots X_n)^{\mathfrak{C}}$ (with $0 \leq n$) where a is a variable and all X_i are SN, then M is SC, and*

2. *every SC term with hat $\mathfrak{C}$ is SN.*

Proof. Proof is by induction on $\mathfrak{C}$.

- $\mathfrak{C}$ is atomic or ☺: since $M^{\mathfrak{C}}$ is of the form $(aX_1 \ldots X_n)^{\mathfrak{C}}$ (with $0 \leq n$) where a is a variable and all X_i are SN, M itself is also SN. (For M to have an infinite reduction sequence, some X_i would need to have one.) But then by Definition 8, M is SC. In this case, clause (2) is immediate from Definition 8.

- $\mathfrak{C}$ is $\rho \to \sigma$:
    - For (1): take any SC term Y^ρ. By IH(2), Y is SN. Now consider $(MY)^\sigma$; by IH(1) this is SC. But Y was arbitrary, so M too is SC by Definition 8.
    - For (2): suppose $M^{\mathfrak{C}}$ is SC, and take some variable x^ρ not occurring at all in M. By IH(1), x is SC. So $(Mx)^\sigma$ is SC too, by Lemma 3. By IH(2), then, Mx is SN as well; and so M is SN by Lemma 5.

- $\mathfrak{C}$ is $\neg \rho$:
    - For (1): take any SC term Y^ρ. By IH(2), Y is SN. Now consider $(MY)^{☺}$; by the base case this is SC. But Y was arbitrary, so M too is SC by Definition 8.

- For (2): suppose $M^{\mathfrak{S}}$ is SC, and take some variable x^ρ not occurring at all in M. By IH(1), x is SC. So $(Mx)^\circledcirc$ is SC too, by Lemma 3. By Definition 8, then, Mx is SN as well; and so M is SN by Lemma 5. □

It follows immediately from Lemma 7 that all variables are SC; I'll appeal to this a few times as we go on.

Lemma 8 (q.v. Hindley & Seldin, 2008, Lemma A3.11, p. 295) *For any types ρ and σ,*

1. *if $M^\sigma[x^\rho \mapsto N^\rho]$ is SC, and if either N^ρ is SC or x is free in M, then $((\lambda x.M)^{\rho \to \sigma} N^\rho)^\sigma$ is SC;*

2. *if $M^\circledcirc[x^\rho \mapsto N^\rho]$ is SC and x occurs free in M, then $((\lambda x.M)^{\rho \to \sigma} N^\rho)^\sigma$ is SC; and*

3. *if $M^\circledcirc[x^\rho \mapsto N^\rho]$ is SC and x occurs free in M, then $((\lambda x.M)^{\neg \rho} N^\rho)^\circledcirc$ is SC.*

Proof.

1. Recall that σ is $\tau_1 \to \ldots \to \tau_n \to \theta$, where θ is either atomic or $\neg \psi$.

   (a) If θ is atomic, take any SC $M_1^{\tau_1}, \ldots, M_n^{\tau_n}$. Since $M[x \mapsto N]$ is SC, it follows from Lemma 4 that $\bigl((M[x \mapsto N])M_1 \ldots M_n\bigr)^\theta$ is SC, and so (since θ is atomic) SN.

   Since this term is SN, by Lemma 5 so are all its subterms, among them $M[x \mapsto N]$ and all the M_is. By Lemma 6, M itself is SN. If x occurs free in M, then N is a subterm of $M[x \mapsto N]$, so N is SN. If x does not occur free in M, then we have by assumption than N is SC, so by Lemma 7 N is still SN.

   Now, suppose toward a contradiction that $\bigl((\lambda x.M)NM_1 \ldots M_n\bigr)^\theta$ is not SN, that it has an infinite reduction sequence. This sequence cannot consist entirely of reductions within $M, N, M_1, \ldots, M_n$, since we know all of these are SN.

   So it must look like this:

   $$(\lambda x.M)NM_1 \ldots M_n \triangleright_\beta (\lambda x.M')N'M'_1 \ldots M'_n$$
   $$\triangleright_{1\beta} (M'[x \mapsto N'])M'_1 \ldots M'_n$$
   $$\triangleright_\beta \ldots$$

By Lemma 2, however, $M[x \mapsto N] \triangleright_\beta M'[x \mapsto N']$,[11] and so we can construct an infinite reduction sequence

$$(M[x \mapsto N])M_1 \ldots M_n \triangleright_\beta (M'[x \mapsto N'])M'_1 \ldots M'_n$$
$$\triangleright_\beta \ldots$$

But this is impossible, since we know $(M[x \mapsto N])M_1 \ldots M_n$ is SN. Thus, $\bigl((\lambda x.M)N M_1 \ldots M_n\bigr)^\theta$ is itself SN. Since θ is atomic, this means $\bigl((\lambda x.M)N M_1 \ldots M_n\bigr)^\theta$ is SC. Since $M_1, \ldots, M_n$ were arbitrary, it follows by Lemma 4 that $(\lambda x.M)N$ is SC.

(b) If θ is $\neg\psi$, take any SC $M_1^{\tau_1}, \ldots, M_n^{\tau_n}, O^\psi$. Since $M[x \mapsto N]$ is SC, it follows from Lemma 4 that $\bigl((M[x \mapsto N])M_1 \ldots M_n O\bigr)^\ominus$ is SC, and so by Definition 8 SN as well.

From here, the reasoning continues just as for the atomic case, but with O hanging out on the right.

2. The reasoning is exactly parallel to the previous case, except slightly simpler, since we are now sure x occurs free in M.

3. The reasoning is a simplified version of the previous cases. Since $M^\ominus[x^\rho \mapsto N^\rho]$ is SC, it follows from Definition 8 that is is SN as well. From this, it follows by Lemma 6 that $M^\ominus$ is SN, and by Lemma 5 that N is SN.

Now, suppose towards a contradiction that $(\lambda x.M)^{\neg\rho}N$ is not SN, that it has an infinite reduction sequence. This sequence cannot consist entirely of reductions in M and N, since these are both SN. So it must look like this:

$$(\lambda x.M)^{\neg\rho}N \triangleright_\beta (\lambda x.M')^{\neg\rho}N'$$
$$\triangleright_{1\beta} M'[x \mapsto N']$$
$$\triangleright_\beta \ldots$$

where $M \triangleright_\beta M'$ and $N \triangleright_\beta N'$.

---

[11] We can see that $M'[x \mapsto N']$ is defined, as the Lemma requires, since it occurs as a subterm in the reduction sequence.

By Lemma 2, however, $M[x \mapsto N] \triangleright_\beta M'[x \mapsto N']$, and so we can construct an infinite reduction sequence:

$$M[x \mapsto N] \triangleright_\beta M'[x \mapsto N']$$
$$\triangleright_\beta \ldots$$

But this is impossible since we know that $M[x \mapsto N]$ is SN. Thus, $((\lambda x.M)^{\neg \rho} N^\rho)^\odot$ is itself SN; since it has hat $\odot$, this makes it SC as well. $\square$

Lemma 9 (q.v. Hindley & Seldin, 2008, Lemma A3.12, p. 296) *For every term M, for all $x_1^{\rho_1}, \ldots, x_n^{\rho_n}$ (with $n \geq 1$), and all SC terms $N_1^{\rho_1}, \ldots, N_n^{\rho_n}$ such that for all $2 \leq i \leq n$ none of $x_1, \ldots, x_{i-1}$ occurs free in N_i, the term $M^\star := M[x_n \mapsto N_n] \ldots [x_1 \mapsto N_1]$ is SC.*

Proof. Induction on the formation of M.

- If M is x_i for some i, then $M^\star$ is N_i. (This relies on the non-freedom assumption if $i > 1$.) By assumption, then, $M^\star$ is SC.

- If M is some other variable, then $M^\star$ is that variable too. By Lemma 7, then, $M^\star$ is SC.

- If M is $M_1 M_2$, then $M^\star$ is $M_1^\star M_2^\star$. By the induction hypothesis, $M_1^\star$ and $M_2^\star$ are both SC. So $M^\star$ is SC by Lemma 3.

- If M is $\lambda x^\rho.N^\sigma$, first choose bound variables in M so that x^ρ does not occur free in any of $N_1, \ldots, N_n, x_1, \ldots, x_n$. Then $M^\star$ is $\lambda x^\rho.N^\star$. By Definition 8, we can show that $M^\star$ is SC by showing that for all SC O^ρ the term $M^\star O$ is SC. So take any SC O^ρ. Then $N[x_n \mapsto N_n] \ldots [x_1 \mapsto N_1][x \mapsto O]$ is SC by the inductive hypothesis applied to N with the sequence $O, N_1, \ldots, N_n$. But this is $N^\star[x \mapsto O]$. So by Lemma 8, $(\lambda x.N^\star)O$ is SC; this is $M^\star O$, so $M^\star$ is indeed SC. $\square$

Theorem 3 *Every term is SN.*

Proof. Take any term M and variable x. By Lemma 9, $M[x \mapsto x]$ is SC; but this is just M itself, so M is SC. By Lemma 7, then, M is SN. $\square$

4 Conclusion

Core type theory is a strange beast. Although core logic is very similar to intuitionistic logic, and core type theory bears a similar relationship to core logic that the simply-typed lambda calculus does to intuitionistic logic, many important features of the simply-typed lambda calculus do not hold of core type theory. Although core type theory is inspired by term rewriting systems like the simply-typed lambda calculus, it is not itself a term rewriting system, since reduction in a subterm can change its context. Core type theory is not confluent, and the equivalence relation generated by its reduction relation is trivial, relating every term to every other.

Taking all this into consideration, it is remarkable that core type theory is as well-behaved as it is. Reduction does not preserve free variables, but it never adds any.[12] Reduction does not preserve type, but it never carries a term of one type to a term of any other type, or an exceptional term to a typed one. Finally, the calculus is strongly normalizing: there are many strange reduction paths, but no infinite ones.

References

Baader, F., & Nipkow, T. (1998). *Term Rewriting and All That*. Cambridge University Press.

Barthe, G., & Melliès, P.-A. (1996). On the subject reduction property for algebraic type systems. In *International Workshop on Computer Science Logic* (pp. 34–57).

Church, A. (1941). *The Calculi of Lambda-Conversion*. Princeton, New Jersey: Princeton University Press.

Girard, J. (1995). Linear logic: Its syntax and semantics. In J. Girard, Y. Lafont, & L. Regnier (Eds.), *Advances in Linear Logic* (pp. 1–42). Cambridge University Press.

Harper, R. (2016). *Practical foundations for programming languages* (2nd ed.). Cambridge University Press.

Hindley, J., & Seldin, J. P. (2008). *Lambda-Calculus and Combinators: an Introduction*. Cambridge: Cambridge University Press.

---

[12]This is exactly as in the simply-typed lambda calculus. Note, however, that in the λI calculus, reduction is more precise: reduction leaves the set of free variables exactly the same. For more on λI, see, e.g., (Church, 1941), where it is called simply 'the calculus of λ-conversion', with what is now more usual marked as 'the calculus of λ-K-conversion'.

Howard, W. H. (1980). The formulae-as-types notion of construction (1969). In *To HB Curry: Essays on combinatory logic, lambda calculus, and formalism*. Academic Press.

Petrolo, M., & Pistone, P. (2019). On paradoxes in normal form. *Topoi*, *38*(3), 605–617.

Scherer, G. (2017). Deciding equivalence with sums and the empty type. In *ACM SIGPLAN Notices* (Vol. 52, pp. 374–386).

Sørensen, M. H., & Urzyczyn, P. (2006). *Lectures on the Curry-Howard isomorphism*. Elsevier.

Tait, W. W. (1967). Intensional interpretations of functionals of finite type I. *The journal of symbolic logic*, *32*(2), 198–212.

Tennant, N. (1979). Entailment and proofs. *Proceedings of the Aristotelian Society*, *79*(1), 167–189+viii.

Tennant, N. (2002). Ultimate normal forms for parallelized natural deductions. *Logic Journal of the IGPL*, *10*(3), 299–337.

Tennant, N. (2015). The relevance of premises to conclusions of core proofs. *Review of Symbolic Logic*, *8*(4), 743–784.

Tennant, N. (2017). *Core Logic*. Oxford: Oxford University Press.

Terese. (2003). *Term Rewriting Systems*. Cambridge University Press.

Troelstra, A. S., & Schwichtenberg, H. (2000). *Basic Proof Theory* (2nd ed.). Cambridge: Cambridge University Press.

David Ripley
Monash University
Australia
E-mail: `davewripley@gmail.com`

Negation on the Neo-Australian Plan

SEBASTIAN SEQUOIAH-GRAYSON[1]

Abstract: A popular contemporary view of negation is that negation may be understood usefully as an intensional connective underpinned by a two-place, *symmetric* compatibility relation. I show here that there are sensible directionally sensitive intensional negations that are underpinned by a *non-symmetric* compatibility relation. This in done by understanding the underpinning compatibility relation to be non-primitive, and by giving an account of it in terms of a binary combination operation and partial ordering.

Keywords: negation, modality, compatibility

1 Introduction

Classical negation is an *extensional* one-place connective insofar as the truth-value of $\sim\!A$ at some point x in a given model **M** depends simply on the truth-value of the negand A at that same point. Classically, $x \vDash \sim\!A \Leftrightarrow x \nvDash A$. By contrast, an *intensional* negation will be one where the truth-value of $\sim\!A$ at some point x in a given model **M** depends on the truth-value of A at some *other* point or points $y, z \ldots$ in the model, such that those points are related to x in some special manner or other. Any one-place intensional connective will be a member of the family of modalities, the most well known of which is the $\Box$-operator from modal logics. Canonically, $x \vDash \Box A \Leftrightarrow \forall y(Rxy \Rightarrow y \vDash A)$. The properties possessed by the two-place accessibility relation R will decide the modal logic in question. For example, if R is an equivalence relation, then $\Box$ will correspond to the necessity operator from S5.

This much at least is old news. Less old is news of the Australian Plan for negation (Berto & Restall, 2019). On the Australian Plan, negation is understood to be a one-place intensional connective, a modality of a particular sort. This negation is introduced via a *compatibility frame*, $\mathbf{F_C} : \langle S, \sqsubseteq, C \rangle$,

---

[1] I would like to thank the audience at *Logica 2019* for their careful attention and valuable feedback. Although there are too many people to thank individually, special thanks must be given to David Makinson, Igor Sedlár, and Andrew Tedder for their invaluable suggestions. I would also like to thank an anonymous referee at *The Logica Yearbook 2019* for their extremely thorough and helpful comments. I am very much in their debt.

where S is a set of information states $x, y, z, \ldots$, $\sqsubseteq$ is a partial order on the members of S, and C is a binary compatibility relation on members of S such that xCy means that the information carried by x is compatible with the information carried by y.[2] C and $\sqsubseteq$ interact as we would expect:

$$\text{If } x \sqsubseteq y \text{ and } yCz, \text{ then } xCz \qquad (1)$$

This makes sense. If the information carried by x is contained within the information carried by y, and the information carried by y is compatible with the information carried by z, then the information carried by x is compatible with the information carried by z also.

A compatibility model $\mathbf{M_C} : \langle \mathbf{F_C}, \Vdash \rangle$, is a pair consisting of our compatibility frame $\mathbf{F_C}$ and the two place relation $\Vdash$, which holds between members of S and formulas. Writing $x, y \ldots \in \mathbf{F_C}$ as an abbreviation for $x, y \ldots \in S$ where $S \in \mathbf{F_C}$, we read $x \Vdash A$ as "x satisfies A", or "the information state x carries information of type A". The precise reading will depend on the application for which the model is being designed, on the domain of S itself. Satisfaction is preserved upwards for atomic propositions p, such that if $x \Vdash p$ and $x \sqsubseteq y$, then $y \Vdash p$.

Things are still rather abstract at this stage, but we will look at specific applications in detail in the sections below. For now, note that we can use our compatibility relation to get the following model theoretic condition for our intensional negation $\sim A$:

$$x \Vdash \sim A \text{ iff } \forall y \in \mathbf{F_c} \text{ s.t. } xCy, y \nVdash A \qquad (2)$$

This makes sense as well. x will carry the information that $\sim A$ just in case for any information state y such that x is compatible with y, y does not carry the information that A.

Just as with the two-place accessibility relation R from modal logics, we may assign various properties to the two-place compatibility relation C. This is not to say that such an assignment will be unconstrained. Given the natural

---

[2]The same negation may be introduced via an *incompatibility*, or perp frame $\langle S, \sqsubseteq, \perp \rangle$, where $\sqsubseteq$ is as before, and $\perp$ is a binary incompatibility relation on members of S (Dunn, 1993). Now $\sim Y := \{X : Y \perp X\}$, with $x \perp y$ being read as *the information carried by* x *is incompatible with the information carried by* y. Translation between compatibility frames and perp frames is straightforward on account of $\forall x, y((xCy) \Leftrightarrow (x \not\perp y))$. Although the entire argument given above and below can be given in terms of incompatibility just as well as it can be given in terms of compatibility, we will restrict ourselves to the latter. Uses of '(in)compatibility' above serve merely to emphasise this former fact.

meaning of compatibility, some properties may be more sensible than others. In particular, what should we say sensibly about symmetry on compatibility?

According to the Australian plan for negation, *(in)compatibility is symmetric* (Berto, 2015; Berto & Restall, 2019; Restall, 1999).

> Whatever kinds of entities a and b are, it seems that if a is (in)compatible with b then b has to be (in)compatible with a (if a's obtaining rules out that of b, then b's obtaining rules out that of a, etc). (Berto & Restall, 2019, p. 21)

According to the neo-Australian plan for negation that I shall be proposing and defending here, *(in)compatibility is non-symmetric* (neither symmetric nor asymmetric). There are cases where a *is* (in)compatible with b, but b is *not* (in)compatible with a.

2 Negation as ruling out

To motivate the neo-Australian plan, note that negations of any type, be they extensional or intensional, will have it in common that, at the very least, they are ruling *something* out (see Berto & Restall, 2019, p. 4).[3] Consider the familiar Boolean negation from classical propositional logic. Such a negation is *ruling out truth*. Indeed, this is exactly what it is that is specified by the semantic clause for classical negation, namely that $x \vDash {\sim} A \Leftrightarrow x \nvDash A$. In the classical case, the ruling out of truth will imply falsity by definition. However, note that the mere act of the ruling out of truth alone does not imply this in general, consider constructive negations originating from intuitionistic logic for example.

What classical and constructive negations do have in common is that they are *static*. They do not range over, that is rule out, dynamic processes, procedures, or actions. By contrast, a *dynamic* negation *will* rule out such processes, procedures, or actions, whatever these might be.[4] It is by exploring dynamic negation in detail that the non-symmetry of compatibility becomes apparent. To bring the details of dynamic negations into focus, we need to introduce an *operationalised information frame*.

---

[3] See also (Sequoiah-Grayson, 2009), and (Sequoiah-Grayson, 2010) for elaborations on the theme of negation as ruling out, or as *prohibition*.

[4] Consider the pragmatic context within which one performs the speech act of either asserting or denying a claim (Restall, 2013). Such an act is not to merely rule out the truth of the proposition being denied, but to rule out the assertion of that same proposition. Similarly, the speech act of assertion rules out the act of denial.

Sebastian Sequoiah-Grayson

An information frame $\mathbf{F_I}$, is a triple $\langle S, \sqsubseteq, \bullet \rangle$, where S and $\sqsubseteq$ are as before, and $\bullet$ is a binary composition operator on members of S. The properties of the composition operator $\bullet$ are a matter of choice. Whether or not the choice is a sensible one depends upon the application in question. For example, we might want $\bullet$ to commute, such that $\forall x, y((x \bullet y \sqsubseteq z) = (y \bullet x \sqsubseteq z))$, in which case we would have a *commuting information frame*. As it happens however, for our purposes of demonstrating sensible applications of an intensional negation built upon a non-symmetric compatibility relation, we want an information frame that is *non-commuting*.

We introduce a non-commuting frame $\mathbf{F_{nc}}$. $\mathbf{F_{nc}}$ is a triple $\langle S, \sqsubseteq, \bullet \rangle$ where S and $\sqsubseteq$ are as above, and $\bullet$ is a *non-commuting* binary composition relation on members of S such that $\exists x, y((x \bullet y) \neq (y \bullet x))$. With this non-commuting frame, we can get not one, but two directionally-sensitive *dynamic* implications—a split-implication pair $\langle \rightarrow, \leftarrow \rangle$. This is done as follows:

$$x \Vdash A \rightarrow B \text{ iff } \forall y, z \in \mathbf{F_{nc}} \text{ s.t. } x \bullet y \sqsubseteq z, \text{ if } y \Vdash A \text{ then } z \Vdash B \quad (3)$$

$$x \Vdash B \leftarrow A \text{ iff } \forall y, z \in \mathbf{F_{nc}} \text{ s.t. } y \bullet x \sqsubseteq z, \text{ if } y \Vdash A \text{ then } z \Vdash B \quad (4)$$

Our split implication pair is a pair of dynamic implications precisely because their model conditions, (3) and (4) make reference to explicit combinatorial operations on the pair of information states x and y.

In what follows we will identify information states strictly with the propositions that they are taken to support. Take (3) as an example. The model condition states that if you combine the information state that satisfies the left-to-right conditional, $A \rightarrow B$, namely x, with the information state that satisfies A, namely y, the result is the information state that satisfies B, namely z. Such informational combination is a dynamic process, and this process is built into the meaning of the conditional itself. The satisfaction condition for the right-to-left conditional in (4) is nearly the same as that for (3), with the difference being that the order of combination for the information states x and y is reversed. The directions of the arrows themselves mark the direction of the combination of the underlying information states. We can think of $A \rightarrow B$ as telling as that the information state underlying A must be combined with the right hand side of the information state underlying the conditional in order to result in B. Similarly, we can think of $A \leftarrow B$ as telling us that the information state underlying B must be combined with the left hand side of the information state underlying the conditional in order to result in B. Just what such conditionals might mean exactly will depend on the domain of S, and we will look at concrete examples in the following two sections.

In order to do this, we need to get our dynamic negation up and running. Our next to final step is to introduce a bottom constant $\mathbf{0}$ in the usual manner such that:
$$x \Vdash \mathbf{0} \text{ for no } x \in \mathbf{F_{nc}} \tag{5}$$
In other words, $\mathbf{0}$ is satisfied by no information state in S.

We can now introduce a directionally sensitive, split dynamic negation pair $\langle A^{\mathbf{0}}, {}^{\mathbf{0}}A \rangle$, the members of which we define in terms of our split implication pair and bottom element, as $A^{\mathbf{0}} := A \to \mathbf{0}$ and ${}^{\mathbf{0}}A := \mathbf{0} \leftarrow A$ respectively. Given these definitions, the satisfaction conditions for our split negation pair are as follows:

$$x \Vdash A^{\mathbf{0}} \text{ iff } \forall y, z \in \mathbf{F_{nc}} \text{ s.t. } x \bullet y \sqsubseteq z, \text{ if } y \Vdash A \text{ then } z \Vdash \mathbf{0} \tag{6}$$

$$x \Vdash {}^{\mathbf{0}}A \text{ iff } \forall y, z \in \mathbf{F_{nc}} \text{ s.t. } y \bullet x \sqsubseteq z, \text{ if } y \Vdash A \text{ then } z \Vdash \mathbf{0} \tag{7}$$

Our split negation pair is a pair of dynamic negations precisely because each negation rules out a certain type of dynamic combinatorial procedure. Take (6) as an example. The model condition for $A^{\mathbf{0}}$ states that if you combine the information state that satisfies this negation, $A^{\mathbf{0}}$, namely x, with the information state that satisfies A, namely y, the result is no information state at all, since no z will *ever* satisfy $\mathbf{0}$ (by (5) above). Such a ruling out of informational combination is the ruling out of a dynamic process, and it is this very sense of "ruling out" that is built into the meaning of the negation itself. The situation with the dynamic negation in (7), ${}^{\mathbf{0}}A$, is nearly the same as is that for $A^{\mathbf{0}}$, with the difference being that the direction of the ruled out informational combination is reversed. ${}^{\mathbf{0}}A$ is the type of thing that can never have A combined with *it*.

More syntacticly, $A \to \mathbf{0}$ (that is, $A^{\mathbf{0}}$), is telling us that if we give something of type A to its right hand side, then it will return $\mathbf{0}$. However, since $\mathbf{0}$ cannot be returned by definition, by (5), $A \to \mathbf{0}$ it telling us that it is the type of thing can can never accept anything of type A on its right hand side. Similarly, $\mathbf{0} \leftarrow A$ (that is, ${}^{\mathbf{0}}A$), is telling us that if we give something of type A to its left hand side, then it will return $\mathbf{0}$. Again however, since $\mathbf{0}$ cannot be returned by definition, by (5), $\mathbf{0} \leftarrow A$ it telling us that it is the type of thing can can never accept anything of type A on its left hand side. More functionally, we can think of any arrow as standing for a function along with its inputs and outputs, with the direction of the arrow specifying the direction from which the input must be made.

So far so good, but what might count as concrete examples of such a split negation, and how might we make sense of any of this in terms of our compatibility relation C?

3 What is compatibility?

What does it mean to say that x and y are compatible? We can tell a sensible story about compatibility in terms of $\bullet$ and $\sqsubseteq$. In particular, I would like to propose the following:

$$xCy \text{ iff } \exists z(x \bullet y \sqsubseteq z) \qquad (8)$$

This is a promising account. In its right-to-left hand direction, (8) seems almost truistic. If the combination of x and y results in or is included in z, then x and y must be compatible. The left-to-right hand direction is less obvious, but the case for it is strong nonetheless. Compatibility is, I take it, supposed to be a novel relation. By novel I mean that it marks something beyond the mere exclusion of classical inconsistency. Similarly, the converse relation of incompatibility is, presumably, marking something more than mere classical inconsistency. In a dynamic setting, the compatibility of x and y means that their combination will get you something, and that this something must be greater than the mere sum of its parts.

To say that the combination of x and y, $x \bullet y$, is greater than the sum of its parts is to say that *combination* is an operation more subtle than is mere *aggregation*. The aggregation of a pair of information states x and y (or pair of possible worlds or any other points of evaluation in a logical semantics for that matter), is their conjunction $x \& y$. By way of example, consider the pair of propositions A and $A \to B$. There is a real difference between an operation that merely aggregates this pair on the one hand, with one that combines them in such a way as to generate B. *ipso facto* for any operations on the underpinning states x and y such that $x \Vdash A$ and $y \Vdash A \to B$. A conjunction/aggregation of x and y, $x \& y$, will give some w such that $w \Vdash A \wedge (A \to B)$. By contrast, the combination of x with y, $x \bullet y$ will give us some state z such that $z \Vdash B$. The states w and z are, here, very different things, arrived at in very different ways. We might say that the latter combinatorial way is *generative*, whereas the former is merely *collective*. To combine some things together is to generate something new, whereas to aggregate some things together is to merely collect them together.

Collective or aggregative conjunction is the Boolean operation $\wedge$, while a generative conjunction will be a syntactic analogue to •. The syntactic analogue to • is *fusion*, ⊗. The support conditions for ∧ and ⊗ are as follows:

$$x \Vdash A \wedge B \text{ iff } x \Vdash A \text{ and } x \Vdash B \tag{9}$$

$$x \Vdash A \otimes B \text{ iff } \exists y, z \in \mathbf{F} \text{ s.t. } y \bullet z \sqsubseteq x,\ y \Vdash A \text{ and } z \Vdash B \tag{10}$$

We conflate these operations at our peril. We may conjoin, aggregate, or collect together the pair of propositions C and $A \rightarrow B$, in which case we get $C \wedge (A \rightarrow B)$. We cannot however *combine* them in the manner explained above, as there is no *generative* result from an attempt to do so (*mutatis mutandis* for the relevant information states in supporting roles). Indeed we may conjoin, aggregate, or collect together successfully *any* pair of propositions baring those pairs that are contradictory, whereas the set of genuinely generative pairs is smaller. I do not think it implausible that our target notion of *compatibility* marks something in the vicinity of potentially successful generative combinatorial operations such as those marked by our • operator. That is, if two states x and y are compatible then their combination will be generative. This will imply that non-generative combinations will be *incompatible*. The corollary to (8) will be (where ⊥ is the incompatibility relation):

$$x \perp y \text{ iff } \neg\exists z (x \bullet y \sqsubseteq z) \tag{11}$$

Although (11) is a strong condition, I do not think it too strong given the reasoning above. By understanding compatibility to be a guarantee of generative success, we are making a distinction between compatibility on the one hand, and *coherence* on the other. On the view that I am proposing here, all compatible pairs of information states (and their carried propositions) will be coherent, but not all coherent information states will be compatible. At the propositional level again, the pair of propositions C and $A \rightarrow B$ are coherent, but they are not compatible (hence they are incompatible). There must be a real distinction between compatibility and coherence, otherwise why all the fuss about compatibility in the first place (as taking it either to be a primitive, or as (LINK) below.[5] To this extent I think that the right-to-left hand direction of (8) is both sensible and plausible insofar as we want compatibility to mark something more substantial than the mere avoidance of logical inconsistency.

This is not to say that (8) is the only option for understanding C. Far from it in fact. De and Omori (2018, p. 291) propose the following constraint

---

[5] I am indebted to an anonymous referee for getting me to think more seriously about this than I would have otherwise.

on compatibility (where $\Rightarrow$ is *if-then* in the metalanguage):

$$x \sqsubseteq y \Rightarrow xCy \qquad \text{(LINK)}$$

If the reasoning with regard to (8) above is correct, then (LINK) is false. The truth of the antecedent is insufficient for the truth of the consequent. Suppose that $x \Vdash A$ and $y \Vdash A$. In this case we will have it that $x \sqsubseteq y$ since $\sqsubseteq$ is reflexive (recall that $\sqsubseteq$ is a partial order). However, we do not have it that xCy, at least not if compatibility is supposed to mark anything like the novel, generative notion that we have articulated above. In these same terms, $x \bullet y$ here will give you nothing at all. It will not give you $A \wedge A$ on account of the very real distinction between combination and aggregation noted above. We may generalise the situation: $\sqsubseteq$ is reflexive insofar as an information state is included in itself, hence $\forall x\, x \sqsubseteq x$. However, compatibility is non-reflexive, hence $\exists x \neg xCx$, hence $\exists x \forall y \neg(x \bullet x \sqsubseteq y)$. As an example, consider the case where $x \Vdash A \to B$. In this case there is no y such that $x \bullet x \sqsubseteq y$, since $x \bullet x$ gives us nothing at all. This follows from the observations on generative versus aggregative operation above. Although there *will* be restricted cases where xCx, namely those where x carries logical truths of the form $\alpha \to \alpha$, this will not be true in general.[6]

Along with being non-reflexive, compatibility is non-transitive. That is there are cases where $\exists x \exists y \exists z \neg((xCy \wedge yCz) \to xCz)$, which is just to say that there are cases where $\exists x \exists y \exists z \neg((\exists w(x \bullet y \sqsubseteq w) \wedge \exists u(y \bullet z \sqsubseteq u)) \to \exists t(x \bullet z \sqsubseteq t))$. Such a case will be satisfied in any model where, in spite of the non-negated antecedent $\exists x \exists y \exists z(\exists w(x \bullet y \sqsubseteq w) \wedge \exists u(y \bullet z \sqsubseteq u))$ being satisfied, there is no t such that $x \bullet y \sqsubseteq t$. Consider the situation where $x \Vdash A \to B$, $y \Vdash B \to C$, $w \Vdash A \to C$, $z \Vdash C \to D$, and $u \Vdash B \to D$. In such a case the antecedent will be satisfied, however there will be no t such that $x \bullet z \sqsubseteq t$. Put perhaps more perspicuously at the syntactic level, the attempt to combine (as opposed to merely aggregate) $A \to B$ with $C \to D$ will fail result in any output at all.

Importantly, commutation on $\bullet$ implies symmetry on C: if there is some z that results from $x \bullet y$, then x is compatible with y. Similarly, if there is some z that results from $y \bullet x$, then y is compatible with x. However, the converse does not hold. Symmetry on C does not imply full commutation on $\bullet$. All that compatibility requires is that there is *some* result from the combination

---

[6]The frame condition for arrow composition is $x \Vdash A \to B$ iff $\forall y, z \in \mathbf{F}$ s.t. $x \bullet y \sqsubseteq z$, if $y \Vdash B \to C$ then $z \Vdash A \to C$, where $\mathbf{F}$ is some arbitrary frame. The properties on $\mathbf{F}$, *ipso facto* the properties on $\bullet$ will depend on the sort of considerations that we are making presently.

of the information state on the left hand side of C with information state on the right. Symmetry on compatibility merely preserves this state of affairs when the information states have "switched sides" so to speak. As long as there is *some* information state, not necessarily the *same* information state, resulting from the switch, compatibility will be symmetric. Commutation on • is a stronger condition, since it requires that the same state z will result from both $x \bullet y$ and $y \bullet x$. By logic it follows that non-symmetry on C implies non-commutation on •.

If C is non-symmetric, then we can use it to give satisfaction conditions for our split negation pair as follows:

$$x \Vdash A^0 \textit{ iff } \forall y \in \mathbf{F_C} \text{ s.t. } xCy, \ y \nVdash A. \qquad (12)$$

$$x \Vdash {}^0A \textit{ iff } \forall y \in \mathbf{F_C} \text{ s.t. } yCx, \ y \nVdash A. \qquad (13)$$

Given the relationship between C and • explained above, we have it also that:

$$x \Vdash A^0 \textit{ iff } \forall y \in \mathbf{F_{nc}} \text{ s.t. } x \bullet y \sqsubseteq z, \ y \nVdash A. \qquad (14)$$

$$x \Vdash {}^0A \textit{ iff } \forall y \in \mathbf{F_{nc}} \text{ s.t. } y \bullet x \sqsubseteq z, \ y \nVdash A. \qquad (15)$$

If we had commutation on •, then not only would we have symmetry on C, but our split dynamic implication pair would collapse into a single dynamic implication. If we had a single dynamic implication then our split negation pair would collapse into a single dynamic negation.[7]

We are now in a position to examine some sensible real world examples of split, direction sensitive dynamic negations. The first example is from natural language semantics, and the second example is from epistemic logic.

4 Categorial grammars

Directional information compatibility (or lack of it) is the mark of much of the informational behaviour of natural languages. On the domain of natural

---

[7]Wansing (2001) states that "Obviously, [a split negation pair] coincide[s] if the not implausible assumption is made that incompatibility is a symmetric relation." Although we should take this comment of Wansing's to be equivalent to the claim that non-symmetry on (in)compatibility *is* an implausible assumption, we should take it to suggest that Wansing subscribes to the Australian plan for negation. Wansing's arguments against treating negation as a modality, on account of such a treatment imposing an asymmetry on the satisfaction conditions for negations on the one hand and their negands taken separately on the other, that is, between negative and positive information, is well known. See (Wansing, 1991, 2016) for a careful defense.

languages, in this case English, the compatibility relation is non-symmetric. Take an intransitive verb such as 'hops', and a noun such as 'Alice'. 'Alice hops' is well-formed, whilst 'hops Alice' is not. In this case, 'Alice' is compatible with 'hops', however 'hops' is not compatible with 'Alice'. The compatibility here is directional, insofar is the generation of meaningful, well-formed information is concerned.[8]

We can bring these facts together with operational semantics via *multiple-typing*. Typing is the specification of the functional type of a lexical item. Multiple-typing is a consequence of the recognition that lexical items may correspond to more than one type, that is, to multiple types. By way of example, consider again our intransitive verb 'hops'. Writing $[X : Y]$ for X *is of type* Y, ['hops' : $^0 n$], which is just to say that ['hops' : $\mathbf{0} \leftarrow n$]. This is because if you give 'hops' to the left hand side of a noun, you do not get any informational output. Again, 'hops Alice' is not well-formed. However, ['hops' : $n \rightarrow s$] also. That is, if you give 'hops' to the right hand side of a noun, the you get a sentence as the informational output, but if you give 'hops' to the left hand side of a noun, then you get nothing.

For informational compatibility that runs in the opposite direction, take as an example the adjective 'happy', and the noun 'Alice' again. Here 'happy' is compatible with 'Alice'. This is because if you give 'happy' to the left side of a noun n, you get a complex noun phrase np as the informational output. Hence ['happy' : $np \leftarrow n$].

Is 'Alice' compatible with 'happy'? Yes it is, however this is an instance where mere symmetry on compatibility does not imply commutation. The result of giving 'happy' to the right hand side of 'Alice', 'Alice happy', is obviously distinct from 'happy Alice', so we have commutation failure. However, and perhaps less obviously, 'Alice happy' is not meaningful on its own, although it is well-formed, hence it is not the case that ['happy' : n^0], and hence 'Alice' is compatible with 'happy'. What type of proto-meaning does the complex term 'Alice happy' possess? One type is that 'Alice happy' is a subject phrase that is the type of thing that will return a question q, namely 'Is Alice happy?', as an output if you give the modal verb mv 'is' to its left hand side. That is, ['Alice happy' : $q \leftarrow mv$]. However, ['Alice happy' : mv^0] also, since 'Alice happy is' is not well formed. It is doubtful that the latter complex term could be embedded in any larger well-formed meaningful complex term without stretching English to breaking point.

---

[8]Trivially, it is also the case that we get an asymmetry of incompatibility here, as 'hops' is incompatible with 'Alice, but 'Alice' is not incompatible with 'hops'.

Negation on the Neo-Australian Plan

Some of the most dramatic examples of non-symmetry on C then, are manifested by the already well-known non-commuting properties of natural language. Our next example of non-symmetry on C comes from inference-type epistemic actions.

5 Epistemic logics

Dynamic negations, along with a corresponding failure of compatibility symmetry, find another of their philosophically interesting homes within epistemic logics. In particular, they find a home within fine-grained substructural approaches to modelling epistemic actions. The relevant dynamic negation will, in this case, be ruling out certain epistemic actions themselves. Epistemic actions come in various sorts, including but not limited to observations, announcements, and inferences. Here we will look at inference actions in detail.

Dunn (2015) has given our information frame $\mathbf{F_I} : \langle S, \sqsubseteq, \bullet \rangle$ an interpretation in terms of *informational relevance*. Following Dunn, we interpret information inclusion as information relevance itself, so $x \sqsubseteq y$ is read as *the information carried by* x *is relevant to the information carried by* y.[9] Following on from the arguments given in (Sequoiah-Grayson, 2016), we extend Dunn's interpretation of information frames to the combination operation $\bullet$. In this case, $x \bullet y \sqsubseteq z$ is read as *the combination of the information carried by* x *with the information carried by* y *is relevant to the information carried by* z. We have extended informational relevance from a relationship between information states alone, to a relationship between combinatorial operations on pairs of states on the one hand, and further information states on the other.

The next and crucial step is to give $\mathbf{F_I}$ a robust epistemic interpretation such that the members of S are epistemic states of explicit knowledge, that is, factive attitudinal states directed towards propositions, of some agent α, and $\bullet$ marks the psychological epistemic action of combining the objects of epistemic propositional attitudes.[10] Under this epistemic interpretation, $\sqsubseteq$ has moved from bare information relevance to *epistemic relevance*.[11] Here,

---

[9]Dunn's motivation here is to give an informational interpretation of the semantics for relevance logics. The motivation is well-placed. If the information carried by x is included in the information carried by y, then it will be informationally relevant to y.

[10]A doxastic interpretation would do just as well. Doxastic interpretations carry the risk of suggesting that one is excluding factive epistemic phenomena however, so we will focus on an epistemic interpretation for pragmatic reasons if for no others.

[11]See (Aucher, 2014, 2015; Bílková, Majer, & Peliš, 2015; Sedlár, 2015) for recent sub-

we will find cases where commutation will fail on account of compatibility between epistemic states being non-symmetric.

Suppose that $x \Vdash A \to B$, $y \Vdash B \to C$, and $z \Vdash A \to C$. In this case $x \bullet y \sqsubseteq z$, but $y \bullet x \not\sqsubseteq z$ Unlike the former, the latter epistemic action is not epistemically relevant to α's being in state z at all. $x \bullet y$ is the correct order of combination insofar as cutting q is concerned, whereas $y \bullet x$ is not. Any attempt to combine C with an input that accepts A will be epistemically futile. It is an epistemic action that will remain epistemically irrelevant.

Recall that the operational clause $x \bullet y \sqsubseteq z$ and the compatibility relation C are related as follows - xCy *iff* $\exists z(x \bullet y \sqsubseteq z)$. In our example above, we have it that $\exists z(x \bullet y \sqsubseteq z)$, but we do *not* have it that $\exists z(y \bullet x \sqsubseteq z)$ on account of $y \bullet x \not\sqsubseteq z$. In fact there is no epistemic state resulting from $y \bullet x$ at all. Importantly, it is not the case that $x \bullet y$ will result in some epistemic state of α such that this state carries the information that $(B \to C) \wedge (A \to B)$. To think this would be to confuse mere *epistemic aggregation* with *epistemic integration*.

This is not to suggest that epistemic aggregation is less of an epistemic action that is epistemic composition, merely that they differ in important respects. Consider the case where $K_\alpha A$ and $K_\alpha(A \to B)$. Does it follow that $K_\alpha(A \wedge (A \to B))$? Not in any useful sense where we are using K to specify existent propositional attitudes of α. Although it is true that α has been in a factive attitudinal state that takes A as its target proposition, or its object, and it is true that α has been in another factive attitudinal state that takes the conditional $A \to B$ as *its* object, it does not follow from these facts alone that α has been any attitudinal states, factive or otherwise, which takes the conjunction $A \wedge (A \to B)$ to be its target proposition, or its object. To put the same point differently, it does not follow that α has ever stood in any attitudinal relation at all to a conjunction from the mere fact that α has stood in the same attitudinal relation to each of its conjuncts separately. In order for it to be the case that $K_\alpha(A \wedge (A \to B))$ on the basis of $K_\alpha A$ and $K_\alpha(A \to B)$, α must bring the objects of their knowledge together. That is, they must *aggregate* the objects of their knowledge.

Importantly, the result of such attitudinal aggregation, which in the case above is $K_\alpha(A \wedge (A \to B))$, is insufficient for α to come to know that B, that is, for it to be the case that $K_\alpha B$. In a slogan, just because you know something (or a lot of things for that matter) does not mean that you know

---

structural approaches to epistemic logics in general, and (Sequoiah-Grayson, 2016) for an interpretation of informational relevance in terms of epistemic relevance in particular.

how to do anything useful with it. One's ability to act usefully on the objects of one's knowledge will depend on one's logical acumen. In order for α to know that q on the basis of their knowing that A and that $A \to B$, α must *combine* or *compose together* the objects of their knowledge. Such an epistemic action is an epistemic *integration* rather than a mere epistemic aggregation. It is the generative interaction between or fusion of the relevant objects of knowledge, as opposed to their mere conjoining.

The properties of the action of epistemic integration will be whatever properties the integration operation • must have in order that epistemic progress be preserved. The preservation of such epistemic progress will supervene on the logical form of the propositions that are the objects of the epistemic states or attitudes themselves.[12] There may well be particular instances where the epistemic integration operation will commute. In fact the scenario where $x \Vdash A$ and $y \Vdash B$ is one of these. However, this will not be true in general across the domain of epistemic states. Consider again our example above, where we have it that the relevant epistemic states x, y, z of α are such that $x \Vdash A \to B$, $y \Vdash B \to C$, and $z \Vdash A \to C$. In this case $x \bullet y \sqsubseteq z$, but $y \bullet x \not\sqsubseteq z$. This is on account of the fact that xCy but *not* yCx. Although x carries information of a type that can be combined with information of the type carried by y, x does not carry the type of information that can have information of the type carried by y combined with *it*.[13]

Concretely, and typing on epistemic states or propositional attitudes themselves, it is the case that $[y : (A \to B)^0]$, but it is *not* that case that $[x : (B \to C)^0]$.[14] On the space of propositional attitudes with a combinatorial operation of psychological epistemic actions, such operations will be delicately order-sensitive. So delicate that some reorderings may, as we have seen, block epistemic progress to any further epistemic state whatsoever. Hence the (in)compatibility of such states or attitudes is in certain cases non-symmetric.

---

[12]This is not entirely disanalogous to the considerations bearing on the well-formedness and typing of lexical items in the section on categorial grammars above.

[13]Given that direction is now a distinction with a difference, 'application' might be a better term than 'combination'.

[14]This is on account of the fact that, as we have seen, syntactically speaking we have it that $(A \to B) \otimes (B \to C) \vdash (A \to C)$, whereas we do not have it that $(B \to C) \otimes (A \to B) \vdash (A \to C)$.

6 Conclusion

Understanding the compatibility relation C in terms of $\bullet$ and $\sqsubseteq$ has many advantages, and I hope to have made the case in Section 3 that such an understanding is motivated sensibly and not entirely unnatural. Genuinely generative combinatorial progress underpins a compatibility relation that allows in turn an order-sensitive intensional split-negation pair. Each member of the pair is a modality in virtue of their intensional frame conditions, hence the pair is sympathetic with the Australian plan for negation to this extent. The advantages of such a pair turn on the fact that the underlying non-symmetry allows us to preserve the understanding of negation as "ruling out", where such a ruling out extends to directionally-sensitive contexts. Categorial grammars and inference actions are just two such contexts, and we should expect there to be more.

References

Aucher, G. (2014). Dynamic epistemic logic as a substructural logic. In A. Baltag & S. Smets (Eds.), *Johan van Benthem on Logic and Information Dynamics* (pp. 855–880). Cham: Springer International Publishing.

Aucher, G. (2015). When conditional logic and belief revision meet substructural logics. In *Proceedings of the 2015 International Conference on Defeasible and Ampliative Reasoning* (pp. 2–8). Aachen, DEU: CEUR-WS.org.

Berto, F. (2015). A modality called 'negation'. *Mind*, *124*(495), 761–793.

Berto, F., & Restall, G. (2019). Negation on the Australian plan. *Journal of Philosophical Logic*, *48*(6), 1119–1144.

Bílková, M., Majer, O., & Peliš, M. (2015). Epistemic logics for sceptical agents. *Journal of Logic and Computation*, *26*(6), 1815–1841.

De, M., & Omori, H. (2018). There is more to negation than modality. *Journal of Philosophical Logic*, *47*(2), 281–299.

Dunn, J. M. (1993). Star and perp: Two treatments of negation. *Philosophical Perspectives*, *7*, 331–357.

Dunn, J. M. (2015). The relevance of relevance to relevance logic. In M. Banerjee & S. N. Krishna (Eds.), *Logic and Its Applications. Lecture Notes in Computer Science* (Vol. 8923, pp. 11–29). Berlin, Heidelberg: Springer-Verlag.

Restall, G. (1999). Negation in relevant logics (How I stopped worrying and learned to love the Routley star). In D. M. Gabbay & H. Wansing (Eds.), *What is Negation?* (pp. 53–76). Dordrecht: Springer Netherlands.

Restall, G. (2013). Assertion, denial and non-classical theories. In K. Tanaka, F. Berto, E. Mares, & F. Paoli (Eds.), *Paraconsistency: Logic and Applications* (pp. 81–99). Dordrecht: Springer Netherlands.

Sedlár, I. (2015). Substructural epistemic logics. *Journal of Applied Non-Classical Logics*, 25(3), 256–285.

Sequoiah-Grayson, S. (2009). Dynamic negation and negative information. *The Review of Symbolic Logic*, 2(1), 233–248.

Sequoiah-Grayson, S. (2010). Lambek calculi with 0 and test-failure in DPL. *Linguistic Analysis*(35), 517–532.

Sequoiah-Grayson, S. (2016). Epistemic relevance and epistemic actions. In K. Bimbó (Ed.), *J. Michael Dunn on Information Based Logics* (pp. 133–146). Cham: Springer International Publishing.

Wansing, H. (1991). *The Logic of Information Structures. Lecture Notes in Computer Science (Lecture Notes in Artificial Intelligence)* (Vol. 681). Berlin, Heidelberg: Springer-Verlag.

Wansing, H. (2001). Negation. In L. Goble (Ed.), *The Blackwell Guide to Philosophical Logic* (pp. 415–436).

Wansing, H. (2016). On split negation, strong negation, information, falsification, and verification. In K. Bimbó (Ed.), *J. Michael Dunn on Information Based Logics* (pp. 161–189). Cham: Springer International Publishing.

Sebastian Sequoiah-Grayson
University of Sydney, Department of Philosophy
Australia
E-mail: sequoiah@gmail.com

Alethic Pluralism and Logical Consequence

NICHOLAS J.J. SMITH[1]

Abstract: It has been argued that alethic pluralists—who hold that there are several distinct truth properties—face a problem when it comes to defining validity. Via consideration of the classical concept of logical consequence, and of strategies for defining validity in many-valued logics, this paper proposes two new kinds of solution to the problem.

Keywords: alethic pluralism, truth pluralism, mixed inferences, validity, logical consequence, truth preservation, logical form, many-valued logic

1 Introduction

What does truth consist in: correspondence, coherence, warranted assertibility...? Alethic pluralists think that there is not just one correct answer to this question, but that truth consists in different things for different kinds of claims: e.g., provability for mathematical claims, coherence for ethical claims, and correspondence for empirical claims.[2]

An influential objection to alethic pluralism is that there is no adequate definition of validity (logical consequence) that is compatible with the pluralist position. In this paper I respond to this objection. My interest is not so much in defending alethic pluralism as in elucidating the notion of consequence: seeing what is wrong with the objection reminds us of some points about this notion that are of general importance.

The objection to pluralism was presented by Tappolet (1997):

> Consider the following inference: (1) Wet cats are funny. (2) This cat is wet. *Ergo*, this cat is funny. The validity of an inference

---

[1]Thanks to David Makinson, Dave Ripley, Karel Šebela, Göran Sundholm, the audience at Logica in Hejnice on 25 June 2019, the editors and the anonymous referees for helpful comments.
[2]There is a variety of sometimes subtly but significantly different alethic pluralist views. For maps of the literature see, e.g., (Pedersen, 2012) and (Pedersen & Wright, 2013).

> requires that the truth of the premisses necessitates the truth of the conclusion. But how can this inference be valid if we are to suppose... that two different kinds of truth predicates are involved in these premisses? For the conclusion to hold, some unique truth predicate must apply to all three sentences. But what truth predicate is that? And if there is such a truth predicate, why isn't it the only one we need? [¶] Mixed inferences remind us of a central platitude about truth, namely that truth is what is preserved in valid inferences. Moreover, they show that all sentences which can appear in such inferences are assessable in terms of the same truth predicate. The upshot is that only a truth predicate shared by all sentences which can appear in inferences will satisfy the platitude relating truth to inferences. (pp. 209–210)

The problem arises with *mixed inferences*, where at least two of the component propositions are (as the pluralist sees it) in the domains of different truth properties. Some such inferences appear to be valid and some do not. The challenge for the pluralist is to define validity in such a way as to maintain these appearances, without departing (too radically) from the classical understanding of validity as involving necessary truth preservation.

A natural thought for the pluralist is to turn to many-valued logics (MVL) for leads on how to define validity when we have multiple kinds of truth—for the presence of multiple truth values is certainly no object to giving reasonable definitions of validity in MVL. There are at least three standard ways of defining validity in MVL (this list is not exhaustive—these are the most common options): (1) Pick a single one of the truth values and define validity in terms of preservation of that value. (2) Pick a subset of the truth values as *designated* values and define validity in terms of preservation of designatedness. (3) Specify an ordering on the truth values and define validity in terms of that ordering. I shall discuss these options in the order (2)–(3)–(1) in sections 2, 3 and 4 respectively. My discussion of option (2) will be brief because this option has already been proposed in the literature as a model for a pluralist definition of validity—whereas the idea of modelling such a definition on options (1) or (3) is new.

2 Designated values

Beall (2000) writes:

> In the jargon of many-valued logic, validity is to be understood in terms of *designated values*, these being the different ways of being true, as it were. Specifically, an argument is valid iff (necessarily) if all the premises are designated, then the conclusion is designated. (p. 382)

This proposal faces a dilemma however. Designatedness is a *generic* property: it can be possessed by claims from (what the pluralist sees as) different alethic domains (e.g., the component propositions in a mixed inference). If designatedness is a truth property (a kind of truth) then we have no solution to the problem of defining validity for *strong* alethic pluralists, who hold that there are no generic truth properties or that all truth properties are domain-specific. If designatedness is not a kind of truth then we have departed too far from the classical understanding of validity as involving necessary truth preservation.

Rather than discuss these issues further, I shall move on to options (1) and (3)—with the aim of defining validity in a way acceptable even to strong alethic pluralists.

3 Ordering the values

On option (3) we define on ordering on the truth values and say that an argument $\alpha_1, \ldots, \alpha_n / \therefore \beta$ (with premises α_1 through α_n and conclusion β) is valid iff

> (R) there is no model on which the truth value of the conclusion is less than the truth values of all the premises

where 'less than' refers to the ordering. E.g., consider fuzzy logics in which propositions are assigned as truth values real numbers between 0 and 1 inclusive. One standard way of defining fuzzy consequence is to order the truth values in the usual way and apply recipe (R); this kind of consequence relation is sometimes referred to as 'no drop' or 'salvo gradu' consequence.[3]

In the context of alethic pluralism, *how* we might order the truth values (kinds of truth—and falsity) will depend on how many kinds of truth we

---

[3] For further details see (Font, 2003) and (Smith, N. J. J., 2015, pp. 1265–6).

suppose there to be and on why we suppose there to be these multiple kinds of truth (i.e., what the various truth values represent and what work they are supposed to do). Note that nothing in recipe (R) requires the truth values to be ordered in a particular way (e.g., linearly ordered rather than partially ordered). If we have (e.g.) three kinds of truth and one kind of falsity, we might want to (but do not have to) order them in any of the following sorts of way (where an arrow from x to y indicates that x is less than y and the less-than relation is irreflexive and transitive):

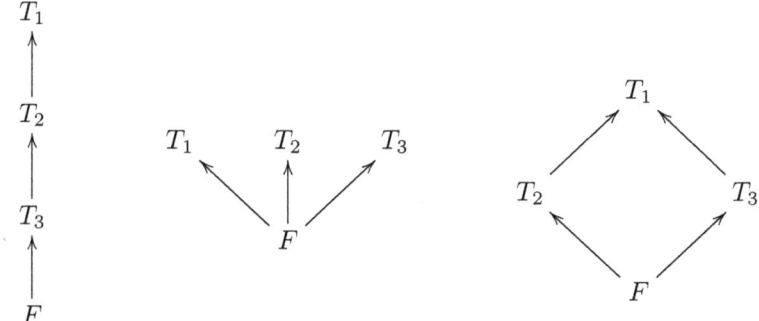

In general, facts about the ordering of the truth values—and facts about how the logical operators behave when given as inputs propositions with various different kinds of truth value—will affect the particular kind of logic that we get when we define consequence in terms of recipe (R). Our concern at present, however, is simply the conceptual task of defining validity. My point is that once we have an ordering on the kinds of truth, we can define validity in terms of preservation of this ordering.[4]

This recipe for defining validity provides a natural implementation of the classical idea of consequence as involving *preservation of truth*. In the classical context, a valid argument has the property that the premisses cannot

---

[4]Cotnoir (2013) (which I discovered after formulating the arguments of this paper) also presents a definition of validity for alethic pluralists that makes use of an ordering on truth values. Cotnoir commits himself to an algebraic semantics in which the truth values are n-tuples, each *component* of which corresponds to one of the kinds of truth countenanced by the pluralist. In my proposal, by contrast, the various truth values simply are (or represent directly) the various kinds of truth and falsity (or in §3.1 below, the various possible truth statuses) and we define an ordering on them directly. Cotnoir's proposal is therefore unnecessarily complex and involves unnecessary additional commitments—e.g., to a certain form of algebraic semantics—and so is dialectically less effective than my proposal: it gives opponents of alethic pluralism additional targets at which to aim. A similar point applies to Pedersen (2006), who presents a definition of validity for alethic pluralists that involves a commitment to plural quantification.

Alethic Pluralism and Logical Consequence

be true without the conclusion being true. There are two interpretations of the idea that this involves 'preserving truth'. One is that there is a single value—truth—such that if all the premises have it then the conclusion has it. The other is that the conclusion is never less true than the premises (where falsity is thought of as less true than truth). Both interpretations are acceptable. They coincide in the classical case—so consideration of that case cannot be used to favour one of them. They can come apart, however, in the context of MVL or alethic pluralism—and my proposal in this section is that pluralists can define a valid argument as one that is truth preserving in the second sense: the conclusion can never be less true than all the premises.[5] This is not to say that the first interpretation of 'preserving truth'—involving a single value that must be preserved—has to be abandoned in the context of alethic pluralism: we shall return to it in §4. Tappolet writes (recall §1) that "only a truth predicate shared by all sentences which can appear in inferences will satisfy the platitude relating truth to inferences". The platitude is that valid inferences preserve truth. Contra Tappolet, it is only on one interpretation of this platitude that there must be a single value or kind of truth that is preserved from premises to conclusion. On a second and equally acceptable interpretation, the core point is that in a valid argument, the conclusion can never be less true than the premises.

3.1 Possessing multiple truth properties

Implicit in the approach of §3 is the assumption that each proposition has exactly one truth value (on each model). What if we have a pluralist view according to which a proposition may possess multiple truth properties at the same time? There are two approaches we can take in this case. One is to pursue a *relational* semantics, in which propositions may be associated with more than one truth value.[6] I shall focus here on a different approach, which is to model a situation in which propositions may possess multiple truth properties using a formal setup in which each proposition possesses exactly one truth value. This kind of approach is generally technically simpler and is widespread in MVL.[7] For example, it is very common to start with the idea that there are two truth values (Truth and Falsity) and some sentences

---

[5] By analogy, we might say that height is preserved across the generations in a given family tree if no child is shorter than both parents, rather than requiring that there be a single particular height that all the generations possess.

[6] Cf., e.g., the relational semantics for FDE in (Priest, 2008, Ch.8).

[7] See (Smith, N. J. J., 2012b) on the distinction between many-valued semantics in the *strict* and *loose* senses.

may possess neither of them, and then model this formally using three truth values—one for each of the three possible truth statuses (as opposed to the two truth values) envisaged in the original motivating story: 'having the value True' (T), 'having the value False' (F) and 'having no value' (N). In this case, a natural ordering of the values is as follows:

Similarly, if one starts with a motivating story that has two truth values and allows not only truth gaps but also gluts, then it is quite standard to model the situation formally using four truth values, one for each of the four envisaged truth statuses: 'having the value True' (T), 'having the value False' (F), 'having no value' (N) and 'having both values' (B). In this case, there are two natural ways of ordering these values (Belnap, 1977)—the truth ordering (on the left) and the knowledge ordering:

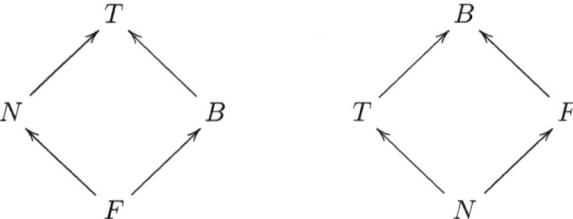

In the case of alethic pluralism, the story might go as follows. We posit (e.g.) three domain-specific truth properties and a generic truth property[8] and allow four possible statuses for propositions: having the first/second/ third domain-specific truth property and the generic truth property ($T_1^g/T_2^g/T_3^g$) or having none of the truth properties (N). This is then modelled using a system that has four truth values—one for each of the four possible truth statuses (not

---

[8] Pluralists who posit a generic truth property have the option of defining validity in terms of preservation of this property—but the present example is merely an example: the general strategy illustrated here is applicable to any pluralist view according to which a proposition may possess multiple truth properties at the same time.

Alethic Pluralism and Logical Consequence

one for each of the four truth properties). Because the four values correspond to the statuses (not the properties) in the original motivating story, it makes perfect sense for each proposition to possess exactly one of them: the statuses are mutually exclusive and jointly exhaustive (while the properties are such that a proposition might possess two of them—or none of them). The story then proceeds as in §3: we order the values and define consequence using (R). A natural ordering of the values might be:

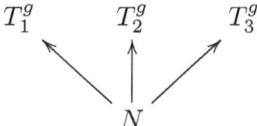

but of course this is just an example—we might want to have a different number of truth statuses/values and we might want to order them in different ways. None of this affects my point about the conceptual task of defining validity in terms of recipe (R), which requires only some truth values and an ordering on them.

4 Defining validity in terms of one value

One might think that a natural option for an alethic pluralist would be to define multiple notions of validity: one for each kind of truth, defined in terms of preservation of that kind of truth. This brings us to option (1): picking one of the truth values and defining validity in terms of preservation of that value—for if we can do this for one of the truth values then we can do it once for each of them (yielding multiple notions of validity, one for each kind of truth). Of course, the pluralist *may* prefer just to single out one notion of truth and define validity in terms of it (rather than having a plurality of kinds of validity to match the plurality of kinds of truth) or to define further notions such as 'valid in all (or some) of the senses corresponding to the kinds of truth *in a specified class*'—where a special case of this would be a notion of 'super validity': truth-preserving for every kind of truth. In any case, the issue now is whether a pluralist can pick a *single* non-generic notion of truth and define validity in terms of preservation of that kind of truth. Tappolet's thought seems to be that the pluralist cannot (recall §1):

> some unique truth predicate must apply to all three sentences....
> Mixed inferences... show that all sentences which can appear

in such inferences are assessable in terms of the same truth predicate.... only a truth predicate shared by all sentences which can appear in inferences will satisfy the platitude relating truth to inferences.

Pedersen (2012) spells out Tappolet's line of thought as follows:

> The inference [(1) If drunk driving is illegal, there are trees. (2) Drunk driving is illegal. ∴ (3) There are trees.]—an instance of modus ponens—is clearly valid [as standardly understood in terms of necessary truth preservation], but the pluralist seems to be unable to account for this. For what property is preserved in the inference? The truth of (2) is given by coherence, but for (3) truth is given by correspondence. So, neither coherence nor correspondence will do. But what property, then, is it? This is the problem of mixed inferences. (pp. 591–2)

What precisely is the argument here? Here's one thought:

> (T1) You cannot define validity in terms of preservation of the value X if there are to be valid arguments in which the component propositions cannot all have the value X.

(T1) is false: witness the classical validity of the argument $A, \neg A / \therefore B$. We define classical validity in terms of preservation of the value True—and here is a valid argument in which the premisses cannot all have this value. However, each premise individually can have the value True—whereas in a mixed inference, there is no notion of truth countenanced by the strong pluralist such that each component proposition, taken individually, may be true in that sense. So perhaps the thought is this:

> (T2) You cannot define validity in terms of preservation of the value X if there is to be a valid argument in which one of the component propositions cannot have the value X.

(T2) is also false: witness the classical validity of $A \wedge \neg A / \therefore B$. We define classical validity in terms of preservation of the value True—and here is a valid argument in which the premise cannot have this value.[9]

---

[9] Someone might still feel that $A \wedge \neg A$ is "in the running" to be true (or truth-apt), it just doesn't get there, so to speak—whereas 'Drunk driving is illegal' is not even in the running (or apt) to be correspondence true. I leave it as a challenge for anyone who finds this thought appealing to try to spell it out clearly and precisely and turn it into a cogent objection to the strategy of the present section for defining validity in the context of alethic pluralism.

Alethic Pluralism and Logical Consequence

The idea that validity involves necessary truth preservation is often spelt out with a conditional: necessarily, if the premises are true, then the conclusion is true. The key point to note here is that—at least as far as the standard classical definition of validity is concerned—this is a material conditional. The idea is that necessarily, it is not the case that the premises are true and the conclusion is not—or it is impossible for the premises but not the conclusion to be true. It is not part of this conception of validity that it must be possible for the premises (all) to be true. On the contrary, if the premises cannot (all) be true, then the material conditional is (necessarily) true. So turning to strong alethic pluralism, the fact that in a mixed inference there is no notion of truth such that the component propositions can all be true in this sense does *not* immediately prevent us defining validity in terms of preservation of truth in this sense.

Of course, a problem looms. If we define validity as necessary preservation of truth of kind X then it seems that *every* inference involving a premise that can only possess truth of some *other* kind will automatically be valid. For example, if we define validity as necessary preservation of correspondence truth, then it seems that not only will Pedersen's argument above be valid, but so will a variant in which we replace premise (1) by 'If there are trees, drunk driving is illegal'—whereas surely this argument should be deemed *invalid*. This brings us to a further key point about the classical notion of validity. Consider these inferences: (A) 'The glass contains water. The glass does not contain H_2O. ∴ The glass contains water.' (B) 'The glass contains water. The glass does not contain H_2O. ∴ The glass contains H_2O.' In both cases it is impossible for the premises both to be true (assuming water is necessarily H_2O). This does not however render both arguments classically valid: (A) is valid and (B) is not. This is so even though in (B) it is impossible for the first premise to be true and the conclusion false. But how can this be?—*if*, as many contributors to this debate claim, validity is necessary truth preservation:

> the Tarskian idea that validity is necessary truth-preservation (Beall, 2000, p. 381)

> the classical account of validity, according to which an argument is valid on condition that the truth of the premises necessitates the truth of the conclusion (Tappolet, 2000, p. 383)

> the standard characterization of validity as necessary truth preservation (Cotnoir, 2013, p. 565).

Nicholas J.J. Smith

Well, validity—on the classical conception—is not simply a matter of necessary truth preservation. For an argument to be valid it must be necessarily truth preserving and furthermore this fact must hold in virtue of the *form* of the argument—i.e., it is not something about the subject matter of the argument that ensures that it is necessarily truth preserving (e.g., the premises talk about water and the conclusion talks about H_2O): it is simply the way the argument is put together that guarantees that the premises cannot be true and the conclusion false. Despite a recent tendency—in introductory logic textbooks, and in papers such as those quoted above—to introduce validity in terms of necessary truth preservation (alone), historically it was generally clear that the notion of validity requires more than this: it requires that the argument be necessarily truth preserving thanks to its form or structure. This view can be found in Tarski's seminal discussion of logical consequence, where it is presented as the traditional, intuitive conception:

> I emphasize...that the proposed treatment of the concept of consequence makes no very high claim to complete originality. The ideas involved in this treatment will certainly seem to be something well known...Certain considerations of an intuitive nature will form our starting-point. Consider any class K of sentences and a sentence X which follows from the sentences of this class. From an intuitive standpoint it can never happen that both the class K consists only of true sentences and the sentence X is false. Moreover, since we are concerned here with the concept of logical, i.e., *formal*, consequence, and thus with a relation which is to be uniquely determined by the form of the sentences between which it holds, this relation cannot be influenced in any way by empirical knowledge, and in particular by knowledge of the objects to which the sentence X or the sentences of the class K refer...The two circumstances just indicated...seem to be very characteristic and essential for the proper concept of consequence...(Tarski, 1956, pp. 414–5)

and indeed the idea goes back to Aristotle.[10] Once we are clear that validity is a matter of necessary truth preservation *in virtue of form*, the apparent problem posed by arguments (A) and (B) disappears. Both arguments are necessarily truth preserving—it is impossible for the premises to be true and the conclusion false—but only argument (A) is so in virtue of its form;

---

[10]For further discussion and references see (Smith, N. J. J., 2012a, Ch.1, §1.4).

Alethic Pluralism and Logical Consequence

hence only (A) is valid.

The lesson carries over to the case of alethic pluralism and mixed inferences. Suppose we take a single kind of truth T_i—from the many countenanced by the pluralist—and define validity as follows: an argument is valid iff

>(P) in virtue of the form of the argument, it is impossible for the premisses all to be T_i while the conclusion is not T_i.

The fact that there can be arguments—mixed inferences—whose component propositions cannot (all) be T_i does not pose a problem for this definition. Some such arguments will be valid—those where the *form* of the argument guarantees that it is impossible to make all the premisses T_i while not making the conclusion T_i—and some of them will not be valid—those where this *is* possible *and* those where it is not possible *but* this impossibility holds in virtue of the particular content of the premisses and conclusion rather than in virtue of the form of the argument. For example, if we define validity as preservation of correspondence truth in virtue of form, then Pedersen's example argument is valid (the impossibility of the premisses but not the conclusion being correspondence true holds in virtue of the form of the argument, given that the form is 'If A then B, $A/ \therefore B$' and assuming that a correspondence true conditional cannot have a correspondence true antecedent without having a correspondence true consequent) while my variant of his example is not (the impossibility of the premisses but not the conclusion being correspondence true holds not in virtue of the form of the argument but in virtue of the content of the second premise—in particular its making a claim that falls in the domain of coherence truth).

Consider Tappolet's original example of the wet cats. The reason we think this argument is valid is because of its form: 'All A's that are B are C. This A is B. $\therefore$ This A is C.' In virtue of its form, it is impossible to make the premisses true without making the conclusion true. This holds whatever we mean by 'true'—assuming only that predication and quantification interact with truth in standard ways. It also holds whatever we put in for A, B and C. Thus, in particular, it holds even if we (a) employ a particular sense of 'true' (one out of the many countenanced by the pluralist) and (b) substitute for A, B and C in such a way that the premisses and the conclusion cannot all be true in this sense. The argument was already valid, in virtue of its form: putting in particular premisses with particular contents will not change this fact.[11]

---

[11] One contribution to this literature which does mention the idea that validity has something to

157

Nicholas J.J. Smith

So, pluralists can define a notion of validity for any notion of truth that they countenance: validity$_i$ is a matter of necessary preservation of truth$_i$ in virtue of form. Whether an argument is valid$_i$ depends on how the logical operators interact with truth$_i$. If the logical operators behave in the same ways with respect to two notions of truth (e.g., $\alpha \wedge \beta$ is T_1 iff α and β are both T_1, and likewise $\alpha \wedge \beta$ is T_2 iff α and β are both T_2, etc) then the two corresponding notions of validity will coincide (extensionally). If the logical operators do not behave in the same ways with respect to two notions of truth (e.g., $\alpha \wedge \beta$ is T_1 iff α and β are both T_1, but $\alpha \wedge \beta$ is T_2 iff either α and β are both T_2 or one of them is T_1 and the other is T_2) then the two corresponding notions of validity might not coincide. (To flesh out the example a little further: if we have two kinds of truth, T_1 and T_2, and two kinds of falsity, F_1 and F_2, and a conjunction operation $\wedge$ that interacts with them as follows:

$\wedge$	T_1	F_1	T_2	F_2
T_1	T_1	F_1	T_2	F_2
F_1	F_1	F_1	F_2	F_2
T_2	T_2	F_2	T_2	F_2
F_2	F_2	F_2	F_2	F_2

then the argument $\alpha \wedge \beta \, / \therefore \alpha$ will be valid$_1$, i.e., necessarily T_1 preserving in virtue of form—because the only way that $\alpha \wedge \beta$ can have the value T_1 is if α and β both have the value T_1—but not valid$_2$, because it is possible for $\alpha \wedge \beta$ to have the value T_2 while α does not have the value T_2: if α is T_1

---

do with logical form is (Cotnoir, 2013)—but there are several problems with his discussion. First, Cotnoir's position is unclear. As quoted above, he first invokes "the standard characterization of validity as necessary truth preservation" (p. 565). He later notes that "logical consequence is a *formal* notion. Validity in formal logic is *independent* of content" (p. 573). He then continues: "Pluralists (even strong pluralists) are not barred from thinking that valid inference depends only on the logical *form* of an argument, and not on the content of the particular premises of an instance of an argument form." In fact, not only are they not barred from this—they (like everyone else) are required to think it, if they want to conform to the classical/Tarskian conception of validity. Second, Cotnoir's comments about form are brief and occur entirely within the context of responding to a particular objection to his algebraic definition of validity (discussed in n.4). Although he says that "arguments with the right sort of formal structure are valid *regardless* of whether the premises are interpreted as being from the same domain or entirely different domains. Whether an inference is 'mixed' or not has no effect on the question of its validity" (p. 573), Cotnoir does not see that this sort of point leads to a stand-alone response to the problem of defining validity for alethic pluralism: i.e., the kind of response given in §4 of this paper—which is distinct from and independent of the response given in §3. Third, Cotnoir does not appreciate that the mixed nature of an inference *can* in certain circumstances show up at the level of form, and hence (contra the claim just quoted) have an effect on the question of validity: see §4.1 below.

and β is T_2 then $\alpha \wedge \beta$ is T_2.) This is not the place to explore such options further: the details will depend on how many notions of truth a pluralist countenances and on why these are countenanced (i.e., what the various truth values represent and what work they are supposed to do).[12]

4.1 Mixed inferences in virtue of form

So far we have considered mixed inferences of the kind mentioned in the literature—e.g., Tappolet's and Pedersen's examples. These inferences are mixed (with respect to some notion of truth T_i) in virtue of their *content*. Such inferences cannot pose a problem for the strategy for defining validity presented in §4, because whether an argument is valid (in the sense corresponding to T_i, i.e., necessarily T_i preserving in virtue of form) is a matter of the *form* of the argument. In order to determine whether an argument is valid, we need to look at its form—but once we have abstracted to the level of form, we have left behind the fact that the inference is mixed.

We can however get the fact that an argument is mixed to show up at the level of *form* by introducing certain kinds of logical operators.[13] For example, suppose that we have two kinds of truth, T_1 and T_2, and one kind of falsity, F. Suppose that the operators † and ‡ are defined so that †α only ever takes the values T_1 or F and ‡α only ever takes the values T_2 or F.[14] In that case the argument †A, ‡B/ ∴ C will be both T_1 valid and T_2 valid: it is impossible *in virtue of the form of the argument* for both premises to be T_1 (without the conclusion being T_1) and similarly for T_2.

I take this to be an observation, not an objection to the strategy for defining logical consequence presented in §4. Consider some comparison cases. In fuzzy logic, it is quite standard to define consequence as necessary preservation of truth degree 1 (in virtue of form) and no problem is posed for this definition by the fact that we can define operators (n) such that $(n)\alpha$ has degree of truth n if α does and otherwise has degree of truth 0—so that (e.g.) the argument $(1)A$, $(.5)B$/ ∴ C is then valid. Likewise in classical

---

[12] Another question for the pluralist is how to define soundness. Classically, an argument is sound if it is valid *and* all its premises are in fact true. There are many possibilities for a pluralist definition of soundness: 'valid' in the classical definition can be replaced by any of the notions of validity discussed above and 'true' can be replaced by any of the kinds of truth countenanced by the pluralist (in which case an inference whose premises are mixed with respect to the notion of truth employed will not be sound) or (e.g.) by a disjunction of some or all of them.

[13] Thanks to Dave Ripley for helpful discussion here.

[14] Beyond these facts, the particular details of the definitions of † and ‡ do not matter here.

logic, no problem is posed for the classical definition of consequence as necessary preservation of truth (in virtue of form) by the fact that we can define an operator $\bot$ such that $\bot\alpha$ is false whatever the value of α—so that the argument $\bot A / \therefore B$ is then valid.[15]

5 Conclusion

I promised strategies for defining validity for alethic pluralists and generally useful reminders about the nature of consequence. Let me gather the reminders here. First, consequence certainly has something to do with 'truth preservation', but this has two equally acceptable interpretations, which coincide in the classical context: preserving a single truth value; or preserving height in an ordering. Second, the conditional often used to spell out the idea of truth preservation—necessarily, if the premises are true then the conclusion is true—is treated in practice as a material conditional: the key point is the impossibility of making the premises but not the conclusion true; there is no implication that the premises can all be true. Third, necessary truth preservation is not enough for validity: logical consequence is a matter of necessary truth preservation *in virtue of form*. Once we have these points clearly in view—and once we appreciate that MVL offers other options for defining logical consequence apart from the idea of preserving designated values—it becomes apparent that there are several workable strategies for defining validity in the context of alethic pluralism.

References

Beall, J. (2000). On mixed inferences and pluralism about truth predicates. *Philosophical Quarterly*, 50(200), 380–382.

---

[15]Of course some think that this very example—and/or its close relatives $\bot / \therefore B$ (where $\bot$ here is a nullary operator, rather than a unary operator as above), $A \wedge \neg A / \therefore B$ and $A, \neg A / \therefore B$—is a problem for the classical definition of validity. Such considerations are however orthogonal to my point in §4, which is that the classical idea of defining validity as necessary preservation of a particular truth value X in virtue of form is not automatically vitiated by the positing of multiple kinds of truth (together with the idea that certain sentences can possess only certain kinds of truth). Given that the aim is to maintain the viability of a classical idea, certain kinds of dialetheists and others who *already* have problems with the classical approach will not find that those problems miraculously disappear when that approach is translated to the context of MVL or alethic pluralism—but they are of course free to try to translate their favoured fixes to these contexts.

Belnap, N. D. (1977). A useful four-valued logic. In J. M. Dunn & G. Epstein (Eds.), *Modern Uses of Multiple-Valued Logic* (pp. 8–37). Dordrecht: D. Reidel.

Cotnoir, A. J. (2013). Validity for strong pluralists. *Philosophy and Phenomenological Research, 86*(3), 563–579.

Font, J. M. (2003). An abstract algebraic logic view of some mutiple-valued logics. In M. Fitting & E. Orłowska (Eds.), *Beyond Two: Theory and Applications of Multiple-Valued Logic* (pp. 25–57). Berlin: Springer-Verlag.

Pedersen, N. J. (2006). What can the problem of mixed inferences teach us about alethic pluralism? *The Monist, 89*(1), 102–117.

Pedersen, N. J. (2012). Recent work on alethic pluralism. *Analysis, 72*(3), 588–607.

Pedersen, N. J., & Wright, C. D. (2013). Introduction. In N. J. Pedersen & C. D. Wright (Eds.), *Truth and Pluralism: Current Debates* (pp. 1–18). Oxford: Oxford University Press.

Priest, G. (2008). *An Introduction to Non-Classical Logic: From Ifs to Is* (Second ed.). Cambridge: Cambridge University Press.

Smith, N. J. J. (2012a). *Logic: The Laws of Truth*. Princeton: Princeton University Press.

Smith, N. J. J. (2012b). Many-valued logics. In G. Russell & D. Graff Fara (Eds.), *The Routledge Companion to Philosophy of Language* (pp. 636–51). London: Routledge.

Smith, N. J. J. (2015). Fuzzy logics in theories of vagueness. In P. Cintula, C. Fermüller, & C. Noguera (Eds.), *Handbook of Mathematical Fuzzy Logic* (Vol. 3, pp. 1237–81). London: College Publications (Studies in Logic, Mathematical Logic and Foundations series).

Tappolet, C. (1997). Mixed inferences: A problem for pluralism about truth predicates. *Analysis, 57*(3), 209–210.

Tappolet, C. (2000). Truth pluralism and many-valued logics: A reply to Beall. *Philosophical Quarterly, 50*(200), 382–385.

Tarski, A. (1956). On the concept of logical consequence. In *Logic, Semantics, Metamathematics: Papers from 1923 to 1938*. Oxford: Clarendon Press.

Nicholas J.J. Smith
University of Sydney, Department of Philosophy
Australia
E-mail: nicholas.smith@sydney.edu.au

Proof-Theoretic Semantics and the Interpretation of Atomic Sentences

PRESTON STOVALL[1]

Abstract: This essay addresses one of the open questions of proof-theoretic semantics: how to understand the semantic values of atomic sentences. I embed a revised version of the explanatory proof system of Millson and Straßer (2019) into the proof-theoretic semantics of Francez (2015) and show how to specify (part of) the intended interpretations of atomic sentences on the basis of their occurrences in the premises and conclusions of inferences to and from best explanations.

Keywords: Proof-Theoretic Semantics, Atomic Sentences, Intended Interpretation, Explanation

1 Introduction

Proof-theoretic semantics (PTS) is an approach toward meaning that uses rules for inferring to and from linguistic or logical expressions as a basis for semantic evaluation. To date, however, there has been little discussion of the semantic values of—or the rules that govern—atomic sentences in PTS. Their values are generally either assumed given from outside the proof system (e.g., Francez, Dyckhoff, & Ben-Avi, 2019, and Francez, 2015), or provided by the inferences from their own assumptions as in (Francez, 2017), or stipulated via definitional systems of the sort found in (Prawitz, 1973). Francez (2015, p. 377) considers this one of the open questions within proof-theoretic semantics: how should we understand the proof-theoretic semantic values of the atoms of an interpreted language?

---

[1]This essay forms a triad with (Stovall, 2019) and (Stovall, in press) which collectively represent an attempt to rewrite the first two chapters of my dissertation. I am indebted to many people for taking the time to talk with me and offer comments on this material. I am particularly grateful for feedback from Robert Brandom, Ulf Hlobil, Daniel Kaplan, Jared Millson, Nissim Francez, Jaroslav Peregrin, Ivo Pezlar, Mark Risjord, and Shawn Standefer at various times over the last few years. I would also like to thank two reviewers for this journal, and the participants at the 2019 Logica conference. Work on this article was supported by the joint Lead-Agency research grant between the Austrian Science Foundation (FWF) and the Czech Science Foundation (GAČR), *Inferentialism and Collective Intentionality*, GF17-33808L.

In this essay I provide an answer to that question by showing how to use a revised version of the explanatory proof system of Millson and Straßer (2019) to individuate introduction and elimination rules for atomic sentences, under an intended interpretation, in terms of their roles in explanation under that interpretation. Within the context of (Francez, 2015), this determines a proof-theoretic semantic value for the atoms of an interpreted language, just as a specific assignment of truth conditions to sentences fixes atomic sentence meaning in model-theoretic semantics. While this doubtlessly does not exhaust everything one could mean by 'proof-theoretic meaning', it provides a formally tractable mechanism for thinking about at least one dimension of linguistic meaning: explanatory role.

2 Overview of the proof-theoretic semantics of Francez

2.1 Conventions

Consider a language $\mathscr{L}$ consisting of a countable set of atoms and the recursively defined set of sentences derived from the atoms and the Boolean operators in the usual way. Let uppercase Latin letters (except X and Y) range over sentences of $\mathscr{L}$, lowercase Latin letters range over atoms, uppercase Greek letters range over finite sets of sentences, and X and Y range over finite sets of literals (atoms and their negations). I abuse notation and use metalinguistic variables as illustrating instances in the text of the essay.

For display of logical relations I use a natural deduction system of the sort employed in (Francez, 2015). Rules of introduction and elimination are associated with each logical operator. These are *rules of inference* in that they determine which inferences can be made to and from the logically complex sentences of $\mathscr{L}$: the introduction rules for an operator * specify the conditions under which a sentence having * as a major operator can be inferred to as the conclusion of an inference, and the elimination rules for * specify the conditions under which such a sentence can be inferred from as a premise in an inference. I treat inferences as single-step derivations in this essay, rather than as acts of inferring, and I understand a derivation is an ordered series of applications of these rules of inference displayed as a tree structure. Thus, each application of a rule will count as an *inference* from the premise(s) to the conclusion and a *derivation* of the conclusion from the premises.

The object-language expressions that occur in the premises and conclusions of applications of rules of inference in these derivation systems are sequents of the form $\Gamma : A$, where the antecedent Γ is a set of sentences that encodes assumptions on which the derivation of the succedent A may depend. This presentation is in what Francez calls *logistic* form, and it has the virtue of displaying the background contexts that the application of a rule may depend on (Francez gives Γ more structure, but that is not relevant here). The rule $\to I$, for instance, reads as follows:

$$\frac{\Gamma, A : B}{\Gamma : A \to B} \to I$$

As a natural deduction system (rather than a sequent calculus) elimination rules operate on succedents rather than antecedents, as can be seen with the elimination rules for the conditional and conjunction:

$$\frac{\Gamma : A \quad \Gamma : A \to B}{\Gamma : B} \to E \quad \frac{\Gamma : A \wedge B}{\Gamma : A} \wedge EL \quad \frac{\Gamma : A \wedge B}{\Gamma : B} \wedge ER$$

This notation allows for using the intuitive rules of natural deduction while preserving the expressive power of sequent calculi in the ability to keep track of the context from which some derivable formula is derivable. This will in turn facilitate tracking various features of context that attend explanatory inferences. Francez (2015) examines a variety of classical and non-classical natural deduction systems; nothing I say here turns on adopting any particular one.

2.2 Overview of PTS

In PTS the semantic values of sentences are delimited by their canonical derivations. For logically complex sentences, these are determined by the introduction and elimination rules for the logical operators (cf. Francez, 2015, Def. 1.5.9; the major premise of an inference rule for a logical operator '*' is the premise containing '*').

Definition 1 (Canonical Derivations of Logically Complex Sentences From Open Assumptions) *Let A be a logically complex sentence.*

1. *A derivation for $\Gamma : A$ is I-canonical for A iff it satisfies one of the two following conditions:*

    - *The last rule applied in the derivation is an I-rule for the main operator of A.*

- *The last rule applied in the derivation is an assumption-discharging E-rule, the major premise of which is some B in Γ, and its encompassed sub-derivations are all canonical derivations of A.*

2. *A derivation for $\Gamma, A : B$ is E-canonical for A iff it starts with an application of an E-rule with A as the major premise.*[2]

The second condition for I-canonical derivations ensures that the commutativity and associativity of disjunction count as part of the meaning of disjunctions. The proof-theoretic semantic value for a sentence A at a context Γ can now be defined as follows (cf. Francez, 2015, p. 38):

Definition 2 (I-Canonical and E-Canonical Comprehension)

- *The I-canonical comprehension of A at Γ $=_{def}$ the collection of all I-canonical derivations for A from Γ.*

- *The E-canonical comprehension of A at Γ $=_{def}$ the collection of all E-canonical derivations for A from Γ, A.*

The use of the term 'comprehension' is meant to signal a contrast with the extensional semantics of model theory. Notice that PTS is hyperintensional: the compositionality of derivations ensures that the I-canonical derivations for A will differ from those for $A \wedge A$ (cf. the discussion in Pezlar, 2018).

2.3 PTS and atomic sentences

The definition for comprehension is general with regard to atoms and logically complex sentences. The notion of canonical derivability, however, applies only to logically complex sentences. For atoms we have (recall that lowercase Latin letters range over the atoms of $\mathscr{L}$):

Definition 3 (Canonical Derivations of Atomic Sentences From Open Assumptions)

1. *A derivation for $\Gamma : p$ is I-canonical for p iff the last rule applied in the derivation is an I-rule for p.*

---

[2] It is possible to loosen this definition and allow that a derivation is E-Canonical for a formula A just in case the first rule applied to A as a major premise is an E-rule for A. This would allow that derivations containing inferences prior to the application of an E-rule for A as a major premise would still count as part of A's meaning. My thanks to Nissim Francez for pointing this out.

2. *A derivation for Γ, p : A is E-canonical for p iff it starts with an application of an E-rule with p as the major premise.*³

Once we have rules for introducing and eliminating atoms, this will determine their I-canonical and E-canonical derivations, which in turn will fix their comprehensions.

3 Explanation as a basis for introducing and eliminating atomic sentences

The notion of an *explanation* has some intuitive appeal for fixing (part of) the proof-theoretic semantic values of atomic sentences. Just as there are both introduction and elimination rules, so are there two orders or directions of explanation. On one hand we may keep a sentence fixed and look to see which contexts or circumstances better explain it—e.g., we might ask what explains the existence of Socrates. On the other hand we may consider sentences across different contexts and look to see which other sentences are better explained by it—and so we might ask what the existence of Socrates explains. Whatever else it is to understand what a sentence means, one who can specify what it explains and what explains it will know something of its meaning. One might wish to call this the 'explanatory comprehension' of atoms in PTS, as to distinguish the range of semantic content fixed by these explanations from other sorts of proof-theoretic meaning, but I suppress that here.⁴

Philosophical logicians are beginning to investigate proof-theoretic formalizations of explanatory inference (see Litland, 2017, Millson, Khalifa, & Risjord, 2018, Millson & Straßer, 2019, Poggiolesi, 2016 and Poggiolesi, 2018). Each of Litland (2017), Millson and Straßer (2019), and Poggiolesi (2016) use a proof-theoretic metalanguage containing explanatory inferences in order to provide introduction and elimination rules for object-language talk whose meaning has been historically difficult to render in precise terms: viz., factive and nonfactive ground (Litland), best explanation (Millson and Straßer), and formal explanation (Poggiolesi). Each proof system distinguishes two sorts of rules and the derivations they define: one set of rules

---

³A similar remark holds here as in footnote 2.

⁴One also might adopt an epistemological stance and consider the reasons we have for believing what we do, as opposed to the reasons there are for things being that way, but I will continue to speak in an ontological register.

governing inferences that are explanatory, and the other governing non-explanatory inferences. The latter set of rules delimits the fragment of the language defined by the logical operators. I will follow this approach here. Litland (2017, p. 283) refers to these as explanatory and plain arguments, Poggiolesi (2016, p. 3149), following Aristotle and Bolzano, as proofs-why and proofs-that, and Millson and Straßer (2019, pp. 128 and 135) consider explanatory arguments and a version of Smullyan's (1968) classical logic for the sequent calculus LK.

What is needed is an account that allows us to infer to and from some sort of explanation (grounding, best, formal, etc). Schematically this would be a pair of rules that tell us 1) that when one is entitled to infer A from some context Γ, and certain other conditions are met, then one can infer that Γ is the relevant sort of explanation for A, and 2) that when A is the relevant sort of explanation for something, and certain side conditions are met, we can infer A. That is, where $:\blacktriangleright$ indicates that the antecedent explains (in whatever sense) the succedent, we want to fill in the side conditions in the following:

$$\frac{\Gamma : A \qquad side\, conditions}{\Gamma :\blacktriangleright A} \qquad \frac{side\, conditions \qquad \Gamma, A :\blacktriangleright B}{\Gamma : A}$$

To do so is to specify when one can infer *to* an explanation, and what one can infer *from* an explanation. The system of Millson and Straßer (2019) has the virtue of specifying side conditions for sorting candidate explanations in order to arrive at the best (if any).

4 The explanatory proof theory of Millson and Straßer

In this part of the essay I present a streamlined version of part 4 of (Millson & Straßer, 2019), though I have translated their sequent notation into a logistic natural deduction notation of the sort Francez uses. Millson and Straßer are interested in non-monotonic or defeasible explanations, and they begin with the idea of a defeasible inference, displayed as follows.

$$\Sigma \,|\, \Gamma :_\Theta A$$

There are two key technical devices in play here. First, inferences are evaluated relative to both an antecedent Γ whose content is explanatory and a background Σ of sentences against which an explanation is assessed but which are not playing a role in the explanation. This allows them to distinguish background information, which may not be relevant to an explanation, from that which is doing the explanatory work (consider inferring 'the match is lit' from 'the match is struck' in situations where 'today is Tuesday' is part of the background but is not relevant to the inference). Second, a defeater set Θ keeps track of whether an inference is defeated. Θ is a set of sets of sentences, and the inference

$$\Sigma \mid \Gamma :_\Theta A$$

is defeated anytime $\Sigma \cup \Gamma$ logically derives all of the members of any element of Θ.

Material axioms are given with antecedents and succedents as finite sets of literals and literals, respectively. Millson and Straßer integrate these material axioms with rules for the classical logical operators, and an inference $\Sigma \mid \Gamma :_\Theta A$ is logical, and hence indefeasible, just in case Θ is empty (see Millson & Straßer, 2019, Lemma 4.4). And because any antecedent Γ with logically complex sentences will, against a given background Σ, derive a set of literals, the material consequences of $\Gamma \cup \Sigma$ can be identified with the material consequences of that set. Some of these defeasible inferences will be explanations, and Millson and Straßer specify a method for sorting explanations so as to find the best. On this basis they give an introduction rule for inferring to a best explanation, and thereafter derive an elimination rule. I will sometimes refer to the antecedent of an explanation as the explanans and to the succedent as the explanandum.

Millson and Straßer take the notion of *sturdiness* as the key desideratum for a theory of best explanation. Roughly, an explanation is sturdy when its explanans would remain a good explanation for the explandum even if all of its competitor explanatia were false—understood as adding the negations of the sentences occurring in those competing explanatia to the background against which the explanation is assessed. There may be no such explanation (no inference from an explanans remains good upon supposition of the falsity of its alternatives). There may also be ties (multiple explanations remain undefeated when all of the competing explanatia are false). We use $\Xi_{\langle \Sigma, \Gamma, A \rangle}$ to denote the set of explanations that compete with Γ for best explaining A at Σ (see Definition 9 for the explicit definition of $\Xi_{\langle \Sigma, \Gamma, A \rangle}$). I will generally speak of a single best explanation in what follows, but it is important to remember that ties are possible. As we will see, sturdiness amounts to something like an introduction rule for best explanations.

Unfortunately, the use of $\Xi_{\langle \Sigma, \Gamma, A \rangle}$ as a means of testing for sturdiness admits a class of counterexamples (see Part 5 for discussion). To avoid this problem I will eventually revise their account of sturdiness, but I begin by laying out that account.

4.1 Inferring to a best explanation

Our task is to determine whether an explanation is the best by determining whether it is sturdy. Where $\Sigma \,|\, \Gamma :_\Theta A$ is the explanation in question, consider the class of potential competitor explanations for A at Σ. This is the set of candidate best explanations for A at Σ, denoted $\mathbf{S}_{\langle \Sigma, A \rangle}$ and referred to as an Antecedent Set (cf. Millson & Straßer, 2019, Definition 4.8).

Definition 4 (Antecedent Set, $\mathbf{S}_{\langle \Sigma, A \rangle}$) *For any $\varnothing \subset \Sigma, \Theta, A \subset \mathcal{L}$, let*

$$\mathbf{S}_{\langle \Sigma, A \rangle} =_{def} \{X \text{ such that } \varnothing \subset X \subset Lit \text{ and } \Sigma \,|\, X :_{\Theta \cup A} A \text{ is provable }\}.$$

Notice that the inclusion of A in the defeater set ensures that we consider only material inferences. To have an example to refer to throughout the discussion, suppose that we are looking for the best explanation for the existence of Socrates (A) at some context (Σ). The antecedent set $\mathbf{S}_{\langle \Sigma, A \rangle}$ will include as possible explanations, we may suppose, that his parents met, that they fell in love, that they were married, that a particular zygote (understood *de re*) was formed, etc. Depending on what information is contained in the background Σ, any one of these might be explanatory, though clearly some are better than others.

We arrive at competitor sets in two steps, first by using a principle to restrict $\mathbf{S}_{\langle \Sigma, A \rangle}$ to those that are the simplest, and then by using two principles to compare this restricted class of competitor explanations to Γ. Both steps are preserved in the revised notion of sturdiness introduced below, but the two principles employed at the second step are put to slightly different uses.

At the first step we invoke a principle that rules out those explanations that are more committive than necessary. We do this by removing any sets from $\mathbf{S}_{\langle \Sigma, A \rangle}$ that are supersets of other sets in $\mathbf{S}_{\langle \Sigma, A \rangle}$. The resulting set is denoted $\mathbf{S}^{\downarrow}_{\langle \Sigma, A \rangle}$. More generally (cf. Millson & Straßer, 2019, Definition 4.9; and recall that X and Y range over finite subsets of the literals of $\mathcal{L}$):

Definition 5 (Literal-Set Minimisation, $\mathbf{S}^{\downarrow}$) *For any $\mathbf{S} \subset \mathcal{P}(Lit)$, let*

$$\mathbf{S}^{\downarrow} =_{def} \mathbf{S} \setminus \{X \text{ such that } Y \in \mathbf{S} \text{ and } Y \subset X\}.$$

Continuing with our example of Socrates, suppose that $\mathbf{S}_{\langle \Sigma, A \rangle}$ includes both {a particular zygote was formed} and {a particular zygote was formed, Socrates exists}. In this case $\mathbf{S}^{\downarrow}_{\langle \Sigma, A \rangle}$ will include only the former explanans, on the principle that if it was explanatory at this background then additional information is otiose.

If the first step involves sorting good from bad explanations within $\mathbf{S}_{\langle\Sigma,A\rangle}$ according to a principle of simplicity, and arriving at the literal-set minimisation $\mathbf{S}^{\downarrow}_{\langle\Sigma,A\rangle}$ as a result, the second step invokes two principles for comparing Γ with the elements of $\mathbf{S}^{\downarrow}_{\langle\Sigma,A\rangle}$ so as to arrive at the right class of competing explanans.

The first principle Millson and Straßer enforce at the second step is another principle of simplicity. Intuitively, if an explanation X for A implies something Y that itself explains A, but which does not in turn imply X, then Y is the simpler explanation and so is preferable to X. And so just as literal set-minimalization removed supersets of other sets in $\mathbf{S}_{\langle\Sigma,A\rangle}$ in order to compare only the simplest explanations with Γ, we want to ensure that any explanation X in $\mathbf{S}^{\downarrow}_{\langle\Sigma,A\rangle}$ that is simpler than Γ will undercut Γ as an explanans by defeating the inference from Γ to A. This defeat is ensured by the fact that, when we add the negations of every sentence in X to Σ, the resulting context-cum-explanans is incoherent. But if an explanation in $\mathbf{S}^{\downarrow}_{\langle\Sigma,A\rangle}$ is more complicated than Γ, we want to be sure *not* to add its negations to the background when testing Γ so as not to defeat the inference from Γ by default.

For instance, where Γ is {a particular zygote was formed, Heraclitus exists} and $\mathbf{S}^{\downarrow}_{\langle\Sigma,A\rangle}$ includes {a particular zygote was formed}, we want to add the negation of 'a particular zygote was formed' to Σ and ensure that the explanans Γ is undercut. By contrast, where Γ is {a particular zygote was formed} and $\mathbf{S}^{\downarrow}_{\langle\Sigma,A\rangle}$ includes {a particular zygote was formed, Heraclitus exists}, we *do not* want to add the negations of the sentences in that latter set to Σ.

To enforce this notion of comparative simplicity we first define 'Δ is logically weaker than Γ' as follows (cf. Millson & Straßer, 2019, Definition 4.10; as a definition for *logical* weakness, notice that the absence of a defeater set indicates use of a logical consequence relation):

Definition 6 (Logically Weaker Than, $\Gamma >^{\vdots} \Delta$) *For any $\Gamma, \Delta \subset \mathcal{L}$, let*

$$\Gamma >^{\vdots} \Delta \text{ iff } [\Gamma : \bigwedge \Delta \text{ but not } (\Delta : \bigwedge \Gamma)].$$

We then require $\Xi_{\langle\Sigma,\Gamma,A\rangle}$ (the defeater set) to include the following:

$$\bigcup \{X \in \mathbf{S}^{\downarrow}_{\langle\Sigma,A\rangle} \text{ such that } \Gamma >^{\vdots} X\}$$

That is, we require that the competitor set include any explantia X that are logically weaker than the candidate explanans Γ, meaning that Γ logically derives every formula in X but not vice versa. Thus when X is a subset of $\Xi_{\langle\Sigma,\Gamma,A\rangle}$, the negations of the sentences of $\Xi_{\langle\Sigma,\Gamma,A\rangle}$ will include negations of formulae that are logically derivable from Γ. It follows that $\Sigma \cup \Gamma$ is incoherent, and so the inference to A will be defeated.

The second principle for comparing Γ with the minimal competitor explanations in $\mathbf{S}^{\downarrow}_{\langle\Sigma,A\rangle}$ is a principle for evaluating competitor explanantia for Γ that are *not* logically weaker than Γ and so which should not straightaway defeat the inference to A. For some explanations Y in $\mathbf{S}^{\downarrow}_{\langle\Sigma,A\rangle}$ are either logically stronger than Γ, in the sense that they imply every sentence in Γ without the converse implication holding, or they are neither logically weaker nor logically stronger than Γ. Of these explanations Y that are not logically weaker than Γ, we must ensure that when we consider whether Γ remains a good explanation on the supposition of their falsity that the explanation to Γ is not undercut simply in virtue of the fact that they share some literals in common with the set of literals implied by Γ. And so we must ensure that $\Xi_{\langle\Sigma,\Gamma,A\rangle}$ does not contain any literals that are implied by Γ and which only occur in explanations Y that are not logically weaker than Γ. We do this by taking the union of the set of explanations Y in $\Xi_{\langle\Sigma,\Gamma,A\rangle}$ that are not logically weaker than Γ and removing any literals that are logically derived by Γ.

Returning to the example concerning Socrates, where $\mathbf{S}^{\downarrow}_{\langle\Sigma,A\rangle}$ includes $X = \{$a particular zygote was formed, Heraclitus exists$\}$, and supposing our candidate explanans is now $\{$a particular zygote was formed, Socrates' parents fell in love$\}$, the fact that X is not logically weaker than Γ means that we must remove the literal that X shares in common with Γ ('a particular zygote was formed') and add only the negation of 'Heraclitus exists' to Σ before seeing whether the explanation still holds. And this is the right result: when I consider whether Γ remains a good explanation even when the competing explanans X doesn't hold, I consider whether the formation of the zygote and his parents falling in love would still explain Socrates' existence if Heraclitus hadn't existed, as this is the only information that is new to that competing explanans (the inclusion of the sentence about his parents falling in love is to ensure that X is not already excluded according to Definition 6).

Let the literal consequence-set of Γ, denoted $\mathrm{Cn}_{Lit}(\Gamma)$ be defined as follows (cf. Millson & Straßer, 2019, Definition 4.11; once again note the restriction to the logical fragment of the consequence relation):

Definition 7 (Literal Consequence-Set, $\mathrm{Cn}_{Lit}(\Gamma)$) *For any $\Gamma \subset \mathcal{L}$, let*

$$\mathrm{Cn}_{Lit}(\Gamma) =_{def} \{l \in Lit \text{ such that } \Gamma : l\}$$

PTS and Atomic Sentences

And so we also include the following in $\Xi_{\langle\Sigma,\Gamma,A\rangle}$:

$$\left(\bigcup\{Y \in \mathbf{S}^{\downarrow}_{\langle\Sigma,A\rangle} \text{ such that } \Gamma \not>^{.} Y\}\right) \backslash Cn_{Lit}(\Gamma)$$

By barring from the competitor set $\Xi_{\langle\Sigma,\Gamma,A\rangle}$ any literals that are consequences of Γ and are only had by explanations Y in $\mathbf{S}^{\downarrow}_{\langle\Sigma,A\rangle}$ that are not logically weaker than Γ, we ensure that those competitors don't straightaway defeat the explanation from Γ when we add the negations of every sentence in $\Xi_{\langle\Sigma,\Gamma,A\rangle}$ to the background and test Γ for sturdiness. Taking these three principles together we have the following general definition (cf. Millson & Straßer, 2019, Definition 4.12; this relation is not named there):

Definition 8 ($\mathbf{S} \backslash\backslash \Gamma$) *For any $X, Y \subset Lit$, $\Gamma \subset \mathscr{L}$ and $\mathbf{S} \subset \mathscr{P}(Lit)$, let*

$$\mathbf{S} \backslash\backslash \Gamma =_{def} \bigcup\{X \in \mathbf{S} \text{ s.t. } \Gamma >^{.} X\} \cup \left(\left(\bigcup\{Y \in \mathbf{S} \text{ s.t. } \Gamma \not>^{.} Y\}\right) \backslash Cn_{Lit}(\Gamma)\right)$$

This says that we arrive at $\mathbf{S} \backslash\backslash \Gamma$ by taking the union of 1) every set in $\mathbf{S}$ that is logically weaker than Γ, together with 2) the union of every set in $\mathbf{S}$ that is not logically weaker than Γ, but subtracting any literal that is a logical consequence of Γ. Consider the following (from Millson & Straßer, 2019, p. 143; more examples are given there):

$$\{\{\neg p\},\{q\},\{p,r\}\} \backslash\backslash \{p,q\} = \{q\} \cup (\{\neg p, p, r\} \backslash Cn_{Lit}(\{p,q\})) = \{q, \neg p, r\}$$

This yields the following definition for competitor sets (cf. Millson & Straßer, 2019, Definition 4.13):

Definition 9 (Competitor Set, $\Xi_{\langle\Sigma,\Gamma,A\rangle}$) *For any $\Sigma | \Gamma :_\Theta A$, let the competitor set of its antecedent Γ be*

$$\Xi_{\langle\Sigma,\Gamma,A\rangle} =_{def} \mathbf{S}^{\downarrow}_{\langle\Sigma,A\rangle} \backslash\backslash Cn_{Lit}(\Gamma)$$

This says that the competitor set for Γ is the set of formulae that are the result of taking the union of the minimized sets that individually materially (defeasibly) explain A against the background Σ, and subtracting any formulae that are literal consequences of Γ and which are implied only by elements of $\mathbf{S}^{\downarrow}_{\langle\Sigma,A\rangle}$ that are not logically weaker than Γ.

To test whether Γ is the *best* explanation for A at Σ we then add the negation of all of the elements of this set (denoted $\neg\Xi_{\langle\Sigma,\Gamma,A\rangle}$) to the background and see whether the explanation still holds. If it does, one is entitled to infer that this is the best explanation. As noted above, this sturdiness rule amounts to something like an introduction rule for best explanations (this rule and the next are given at Millson & Straßer, 2019, p. 144):

$$\frac{\Sigma \mid \Gamma :_{\Theta,A} A \quad \Sigma, \neg \Xi_{\langle \Sigma,\Gamma,A \rangle} \mid \Gamma :_{\Theta,A} A}{\Sigma \mid \Gamma :^{\blacktriangleright}_{\Theta,A} A} \; STR$$

Here the black triangle above the defeater set signifies that Γ is the best explanation for A at Σ.

If Sturdiness (*STR*) functions as a rule for inferring *to* a best explanation and is analogous to an introduction rule for best explanation, then Abduction (*ABD*) functions as a rule for inferring something *from* a best explanation and is analogous to an elimination rule for best explanation:

$$\frac{\Sigma \mid \Gamma, A :^{\blacktriangleright}_{\Theta} B \quad \Sigma' \mid \Gamma' :_{\Psi,B} B}{\Sigma, \Sigma' \mid \Gamma, \Gamma' :_{\Theta,\Psi,B} A} \; ABD$$

5 A problem with Millson and Straßer's account, and a proposed solution

Only some best explanations are captured by Millson and Straßer's test for competitors, however. For anytime there are multiple minimal competitors that contain sentences that cannot all be false together, it will happen that the addition of the negations of the sentences in the competitor set to the background will lead to a contradiction. But from a contradiction anything follows, and this means that in such a case all of the members of some element of the defeater set will be derivable, thus defeating the inference. This can happen even when the competitor explanations are much poorer than the one they are competing against.

This can be seen with an example. Suppose we are evaluating the following explanations for the existence of Socrates: 1) a particular zygote was formed; 2) Socrates' parents fell in love and Heraclitus exists; 3) Socrates' parents fell in love and Heraclitus does not exist. It is clear that the first is the best explanation of this trio. But where

PTS and Atomic Sentences

s = Socrates exists. z = a particular zygote was formed

h = Heraclitus exists. f = Socrates' parents fell in love

and using the familiar notation we have:

$\Gamma = \{z\}$ $\Sigma = \varnothing$

$\mathbf{S}^{\downarrow}_{\langle\Sigma,s\rangle} = \{\{z\}, \{f, h\}, \{f, \neg h\}\}$

$\Xi_{\langle\Sigma,\Gamma,s\rangle} = \left(\mathbf{S}^{\downarrow}_{\langle\Sigma,s\rangle} \setminus\!\setminus Cn_{Lit}(\Gamma)\right) = \left(\varnothing \cup (\{z, f, h, \neg h\}\setminus\{z\})\right) = \{f, h, \neg h\}$

And now when we add the negations of the elements of $\Xi_{\langle\Sigma,\Gamma,s\rangle}$ (that is, f, h and $\neg h$) to Σ we have an incoherent set. As a result, the explanation from z to s is defeated.

The problem lies in the way the competitor set $\Xi_{\langle\Sigma,\Gamma,A\rangle}$ is calculated. Millson and Straßer combine the elements of all of the sets in $\mathbf{S}^{\downarrow}_{\langle\Sigma,A\rangle}$ into one set (subtracting the literal consequences from those sets that are not logically weaker than Γ) and then test Γ against the negations of all of these sentences. Instead, we should test whether an explanation remains good on the supposition of the falsity of each of its competitors separately.[5] We do this as follows (with the superscripted R denoting that we have a revised notion of a competitor set).

Definition 10 (Revised Competitor Set, $\Xi^{R}_{\langle\Sigma,\Gamma,A\rangle}$) *For any $\Sigma \mid \Gamma :_{\Theta} A$, let the revised competitor set of its antecedent Γ be*

$$\Xi^{R}_{\langle\Sigma,\Gamma,A\rangle} =_{def} \left\{ X \text{ such that either } \begin{array}{l} X \in \mathbf{S}^{\downarrow}_{\langle\Sigma,A\rangle} \text{ and } \Gamma >^{\cdot} X, \text{ or} \\ \text{there is some } Y \in \mathbf{S}^{\downarrow}_{\langle\Sigma,A\rangle} \text{ such that} \\ \Gamma \not>^{\cdot} Y \text{ and } X = Y\setminus Cn_{Lit}(\Gamma) \end{array} \right\}$$

Whereas $\Xi_{\langle\Sigma,\Gamma,A\rangle}$ is a set of sentences, $\Xi^{R}_{\langle\Sigma,\Gamma,A\rangle}$ is a set of sets of sentences. This preserves the two-step procedure for arriving at the competitors against which an explanation is to be tested, but the two principles appealed to at the second step are now employed separately, on the elements of $\mathbf{S}^{\downarrow}_{\langle\Sigma,A\rangle}$, to define the revised competitor set. Where $\Xi^{R}_{\langle\Sigma,\Gamma,A\rangle} = X_1, \ldots, X_n$, and where $\neg X_i$

---

[5] A more complicated weighting could be given as well, of course; these are early days for proof-theoretic investigation into the conditions for inferring to and from best explanations.

denotes the negation of every sentence in X, we have the following revised notion of sturdiness (*RSTR*):

$$\frac{\Sigma\,|\,\Gamma :_{\Theta,A} A \quad \Sigma, \neg X_1\,|\,\Gamma :_{\Theta,A} A \quad \ldots \quad \Sigma, \neg X_n\,|\,\Gamma :_{\Theta,A} A}{\Sigma\,|\,\Gamma :^{\blacktriangleright}_{\Theta,A} A} \; RSTR$$

This definition delivers the right result with our example above, and using this basis for inferring *to* a best explanation, the rule for abduction does not need adjustment. I thus propose the following:

Definition 11 (Introduction and Elimination Rules for Atomic Sentences)

- *The introduction rules for an atom p are the applications of RSTR where p is the succedent in the conclusion of the application of the rule.*

- *The elimination rules for an atom p are the applications of ABD where p is in the succedent of the major premise of the application of the rule.*

According to Definition 3 in Section 2.3, this determines the set of canonical derivations for the atoms of $\mathscr{L}$, which thereby determines their semantic values according to Definition 2 in Section 2.2.

6 Summary

Proof-theoretic semantics offers a formally precise and philosophically illuminating investigation into areas of linguistic meaning and rule-governed rationality that have been dominated by model-theoretic semantics and representational notions of cognition over the last century. By establishing a beachhead into these areas on the basis of a semantics for atomic sentences that employs a framework for reasoning to and from best explanations, the possibility of productive proof-theoretic interventions into some of these debates is rendered more likely.

References

Francez, N. (2015). *Proof-theoretic semantics: Studies in logic, vol. 57.* London: College Publications.

Francez, N. (2017). On distinguishing proof-theoretic consequence from derivability. *Logique & Analyse, 238*, 151–166.

Francez, N., Dyckhoff, R., & Ben-Avi, G. (2019). Proof-theoretic semantics for subsentential phrases. *Studia Logica*, *94*, 184–231.

Litland, J. (2017). Grounding ground. In K. Bennett & D. W. Zimmerman (Eds.), *Oxford Studies in Metaphysics, vol.10* (pp. 279–316). Oxford: Oxford University Press.

Millson, J., Khalifa, K., & Risjord, M. (2018). Inferentialist-expressivism for explanatory vocabulary. In V. K. O. Beran & L. Koreň (Eds.), *From rules to meanings: New essays in inferentialism* (pp. 155–178). New York: Routledge.

Millson, J., & Straßer, C. (2019). A logic for best explanations. *Journal of applied non-classical logics*, *29*(2), 124–128.

Pezlar, I. (2018). Proof-theoretic semantics and hyperintensionality. *Logique et analyse*, *61*(242), 151–161.

Poggiolesi, F. (2016). On defining the notion of complete and immediate formal grounding. *Synthese*, *193*(10), 3147–3167.

Poggiolesi, F. (2018). On constructing a logic for the notion of complete and immediate formal grounding. *Synthese*, *195*(3), 1231–1254.

Prawitz, D. (1973). Towards a foundation of a general proof theory. In A. J. P. Suppes L. Henkin & G. Moisil (Eds.), *Logic, methodology and philosophy of science IV* (Vol. 74, pp. 155–178). Amsterdam: North-Holland.

Smullyan, R. (1968). *First-order logic*. Berlin: Springer-Verlag.

Stovall, P. (2019). Characterizing generics are material inference tickets: A proof-theoretic analysis. *Inquiry*, published online March 18, 2019, 1–37.

Stovall, P. (in press). Essence as a modality: A proof-theoretic and nominalist analysis. *Philosophers' Imprint*.

Preston Stovall
University of Hradec Králové, Faculty of Philosophy and Social Science
The Czech Republic
E-mail: `preston.stovall@uhk.cz`

Cut-free and Analytic Sequent Calculus of Intuitionistic Epistemic Logic

YOUAN SU[1] AND KATSUHIKO SANO[2]

Abstract: Artemov and Protopopescu (2016) gave a BHK-interpretation of the knowledge operator based on a verification reading of the intuitionistic knowledge. In their intuitionistic epistemic logic **IEL**, $A \supset KA$ and $KA \supset \neg\neg A$ are accepted as axioms. The work of Krupski and Yatmanov (2016) gave **IEL** a sequent calculus, though the subformula property is not satisfied in the calculus. In this paper we propose an analytic and cut-free sequent calculus for **IEL**. Soundness, completeness and finite model property are established for the new sequent calculus. We also show that our calculus enjoys the disjunction property and Craig interpolation theorem.

Keywords: epistemic logic, intuitionistic logic, sequent calculus

1 Introduction

While epistemic logic based on the classical logic has long been studied after its birth, some works on the intuitionistic approach have gained attention. Williamson (1992) studied an intuitionistic epistemic logic, where he has the equivalence of a proposition and the possibility of a knowledge of the proposition, Proietti (2012) distinguished the implicit and explicit knowledge in Kripke semantics for his intuitionistic epistemic logic.

Different from the classical logic, intuitionistic logic does not commit to the principle of bivalence, which states that for an arbitrary proposition its truth value can be either true or false. Instead of that, intuitionistic logic

---

[1] The work of the first author was partially supported by JSPS KAKENHI Grant-in-Aid for Scientific Research (C) Grant Number 19K12113 and Graduate Grant Program of Graduate School of Letters, Hokkaido University.

[2] The work of the second author was partially supported by JSPS KAKENHI Grant-in-Aid for Scientific Research (B) Grant Number 17H02258, JSPS KAKENHI Grant-in-Aid for Scientific Research (C) Grant Number 19K12113 and JSPS Core-to-Core Program (A. Advanced Research Networks).

follow the Brouwer-Heyting-Kolmogorov interpretation, which states how a proof of a complex formula consists of simpler proofs. For example, a proof of $A \supset B$ is a construction such that, when a proof of A is given, a proof of B can be constructed. Then it is reasonable to ask what can be a proper reading of the knowledge operator in the study of intuitionistic epistemic logic.

Artemov and Protopopescu (2016) gave a BHK-reading of the knowledge operator based on a verification reading of the intuitionistic knowledge. According to their interpretation, a proof of a formula KA ("it is known that A"), is the conclusive verification of the existence of a proof of A. Artemov and Protopopescu (2016) suggest that a conclusive verification can be given by an indirect proof or by an authority. In this reading, $A \supset KA$ expresses that, when a proof of A is given, the conclusive verification of the existence of the proof of A can be constructed. Since a proof of A itself can be regarded as the conclusive verification of the existence of a proof of A, Artemov and Protopopescu claim that $A \supset KA$ is valid.[3]

On the contrary, since the verification does not always give a proof, $KA \supset A$ (usually called *factivity* or *reflection*) is not valid. Let us give an example here. As is well-known, the completeness theorem assures that a valid sentence is always provable in a formal system of first order logic. An argument for the validity of A can be counted as a conclusive verification, which assures us a proof of A in the formal system. But the argument for the validity of A does not immediately give us a proof of A in the formal system. This demonstrates the reason why $KA \supset A$ does not hold in terms of BHK-interpretation.

The system intuitionistic epistemic logic **IEL**$^-$ for belief is obtained from adding **K**-axiom $K(A \supset B) \supset (KA \supset KB)$ and *coreflection axiom* $A \supset KA$ into the intuitionistic propositional logic. Furthermore, if we add *intuitionistic*

---

[3]We received one comment from Göran Sundholm during the LOGICA2019, 24–28, June, 2019, Hejnice, Czech Republic. He criticised the validity of $A \supset KA$. First, he suggested that: "When an implicational proposition $A \supset B$ is asserted as an axiom, it allows one to pass also from *assumptions* that A is true to a conclusion that B is true" (cf. Sundholm, 2014, p. 19). For example, $A \supset (B \supset A)$ says that one can pass from the assumption that both A and B are true to a conclusion that A is true. Then, there is no problem to take $A \supset (B \supset A)$ as an axiom.

Furthermore, Sundholm's explanation on KA is that there is a time when the creative subject has evidence for A (cf Sundholm, 2014, p. 16). If we take $A \supset KA$ as an axiom, it should be able to pass from the assumption that A is true to a conclusion that KA is true. However, Sundholm claims that $A \supset KA$ cannot be an axiom, because: "when the proposition A, whose truth is hypothetically assumed is in fact false, of course, at no stage can the Creating Subject come up with a demonstration that A is true" (cf. Sundholm, 2014, p. 19).

reflection axiom $KA \supset \neg\neg A$ (or equivalently $\neg K\bot$) into **IEL$^-$**, we obtain intuitionistic logic **IEL** for knowledge.

The study of **IEL** also casts light on the study of the knowability paradox. The knowability paradox, also known as the Fitch-Church paradox, states that, if we claim the knowability principle: every truth is knowable $A \supset \Diamond KA$ (let $\Diamond B$ denote that it is possible that B), then we are forced to accept the omniscience principle: every truth is known $A \supset KA$ (Fitch, 1963). This paradox is commonly recognized as a threat to Dummett's semantic anti-realism. It is because the semantic anti-realists claim the knowability principle but they do not accept the omniscience principle. However, as Dummett admitted that he had taken some of intuitionistic basic features as a model for an anti-realist view (Dummett, 1978, p. 164), it is reasonable to consider an intuitionistic logic as a basis. In this sense, if we employ BHK-interpretation of KA as above, then $A \supset KA$ becomes valid and the knowability paradox is trivialized.[4]

Artemov and Protopopescu (2016) also proved that their Hilbert systems are sound and complete for their intended Kripke semantics. As far as the authors know, it is not known yet if their Hilbert systems enjoy the finite model property. Furthermore, Krupski and Yatmanov (2016) gave sequent calculi for both **IEL$^-$** and **IEL**. They also specify the computational complexity of **IEL** as PSPACE-complete by a proof-theoretic method in terms of their sequent calculus. It is remarked, however, that this system does not enjoy the subformula property.

We proceed as follows. In Section 2, we discuss syntax, Hilbert system of **IEL** and **IEl$^-$** and semantics. In Section 3, we propose new sequent calculi named as G(**IEL$^-$**) and G(**IEL**). Both of them satisfy the subformula property. Furthermore the soundness theorem and cut-elimination are proved. We also show the disjunction property and Craig interpolation theorem. In Section 4, we show the finite model property of G(**IEL$^-$**) and G(**IEL**), which also implies semantic proofs of cut-elimination theorems.

2 Syntax and Semantics

The set of formulas of the language $\mathscr{L}$ is defined inductively as:

---

[4]In the problem of knowability paradox, $A \supset KA$ cannot be intuitionistically derived from the knowability principle $A \supset \Diamond KA$, but $A \supset \neg\neg KA$ can be intuitionistically derived from the principle (cf. Su & Sano, 2020). Therefore, a question whether we can accept $A \supset \neg\neg KA$ or not becomes one of the main concerns for intuitionists who accept the knowability principle (e.g., Dummett, 2009, DeVidi & Solomon, 2001, etc.).

$$A := p \,|\, \bot \,|\, A \wedge A \,|\, A \vee A \,|\, A \supset A \,|\, KA,$$

where $p \in \mathsf{Prop}$ and Prop is a countably infinite set of propositional variables.

Definition 1 *We define the set $Sub(A)$ of all subformulas of formula A inductively as follows:*

$$\begin{aligned} Sub(p) &:= \{p\}, \\ Sub(\bot) &:= \{\bot\}, \\ Sub(KA) &:= Sub(A) \cup \{KA\}, \\ Sub(A \circ B) &:= Sub(A) \cup Sub(B) \cup \{A \circ B\}, \end{aligned}$$

where $\circ \in \{\wedge, \vee, \supset\}$. Given a set Ξ of formulas, $Sub(\Xi) := \bigcup \{Sub(A) \mid A \in \Xi\}$.

Definition 2 *A Kripke model is tuple $M = (W, \leq, R, V)$ where W is a nonempty set of possible states, $\leq$ is a preorder on W, R is a binary relation on W satisfying the following two conditions: (1) $R \subseteq \leq$ and (2) $\leq; R \subseteq R$, V is a valuation function: $\mathsf{Prop} \to \mathscr{P}(W)$ satisfying the following persistency condition: if $w \leq v$ and $w \in V(p)$, then $v \in V(p)$, for all $w, v \in W$.*

Definition 3 *Define $\mathbb{M}_{\text{all}}$ as the class of all Kripke models, and define $\mathbb{M}_{\text{ser}}$ as the class of models satisfying the seriality condition: for any $w \in W$, there is a $v \in W$ such that wRv.*

Definition 4 *Given a Kripke model $M = (W, \leq, R, V)$, we define a satisfaction relation $M, w \models A$ as follows:*

$$\begin{aligned} M, w &\models p & &\textit{iff} \quad w \in V(p), \\ M, w &\not\models \bot, \\ M, w &\models A \wedge B & &\textit{iff} \quad M, w \models A \text{ and } M, w \models B, \\ M, w &\models A \vee B & &\textit{iff} \quad M.w \models A \text{ or } M, w \models B, \\ M, w &\models A \supset B & &\textit{iff} \quad \text{for all } v \in W, w \leq v \text{ and} \\ & & & \qquad M, v \models A \text{ imply } M, v \models B, \\ M, w &\models KA & &\textit{iff} \quad \text{for all } v \in W, wRv \text{ implies } M, v \models A. \end{aligned}$$

Given a formula A, a class $\mathbb{M}$ of Kripke models, $\mathbb{M} \models A$ means that we have $M, w \models A$ for all models $M \in \mathbb{M}$ and all states w in M.

Next, we show the persistency, the condition $\leq; R \subseteq R$ is needed when a formula is of the form KB.

Proposition 1 (Persistency) *For an arbitrary formula A and all states $w, v \in W$, if $w \leq v$ and $M, w \models A$ then $M, v \models A$.*

A Kripke model of intuitionistic propositional logic captures the notion of idealized mathematician from Brouwer (cf. van Dalen, 2013, p. 164). An idealized mathematician is said to be able to remember every proposition proved by himself along the flow of time, which is reflected by our $\leq$-relation. That is to say, if proposition P is proved at the stage w, then proposition P holds also at stages which are $\leq$-accessible from w.

Let us assume that proofs of the idealized mathematician follow BHK-interpretation. We explain that the Kripke semantics of the knowledge operator makes sense in the terms of the idealized mathematician in the setting of Artemov and Protopopescu (2016).

In an arbitrary state w, an idealized mathematician has both proved and verified propositions at hand. Verified propositions hold also in the future, i.e., at stages which are $\leq$-accessible from w. Moreover, according to the axiom $A \supset KA$ (already justified by BHK-interpretation), a proof always implies its verification. We understand that wRv holds, if and only if, $w \leq v$ holds (this allows us to get $R \subseteq \leq$) and the verified propositions in w becomes the proved propositions in v. Semantically, if KA hold at w, then for any R-successor state v (because of the condition $R \subseteq \leq$, v is also $\leq$-successor), the idealized mathematician must have proved A at v. That is to say, when he has a conclusive verification of the existence of a proof of a proposition, there must be a group of the future states in which a proof of the proposition will be obtained in every state of that group.[5]

Artemov and Protopopescu (2016) provided Hilbert systems of intuitionistic epistemic logic **IEL** and **IEL$^-$** (Table 1).

Fact 1 (Artemov & Protopopescu, 2016) *For any formula A,*

1. *A is provable in $\mathsf{H}(\mathbf{IEL}^-)$ if and only if $\mathbb{M}_{\mathrm{all}} \models A$,*

2. *A is provable in $\mathsf{H}(\mathbf{IEL})$ if and only if $\mathbb{M}_{\mathrm{ser}} \models A$.*

The proof of this fact can be found in (Artemov & Protopopescu, 2016, Theorem 4.4, Theorem 8.3).

---

[5] According to Artemov and Protopopescu if a $\leq$-successor of a stage s represents the "principally (logically) possible" state from s, then an R-successor can be thought of as a "possible" state of verification.

Table 1 Hilbert systems H(**IEL**$^-$) and H(**IEL**).

	Hilbert system H(**IEL**$^-$)
($\wedge$-Ax)	$A_1 \wedge A_2 \supset A_i$ ($i = 1$ or $i = 2$)
	$A \supset (B \supset (A \wedge B))$
($\vee$-Ax)	$A_i \supset A_1 \vee A_2$ ($i = 1$ or $i = 2$)
	$(A \supset C) \supset ((B \supset C) \supset (A \vee B \supset C))$
($\supset$-Ax)	$A \supset (B \supset A)$
	$(A \supset (B \supset C)) \supset ((A \supset B) \supset (A \supset C))$
($\bot$-Ax)	$\bot \supset A$
(K)	$K(A \supset B) \supset (KA \supset KB)$
(CR)	$A \supset KA$
(MP)	From A and $A \supset B$, infer B.
	Hilbert system H(**IEL**)
All the axioms and rules of H(**IEL**$^-$) plus:	
(IR)	$KA \supset \neg\neg A$

3 Sequent Calculi

A *sequent*, denoted by $\Gamma \Rightarrow \Delta$, is a pair of finite multisets of formulas. The formulas Γ are called the *antecedent*s of $\Gamma \Rightarrow \Delta$, instead Δ are called *succedent*s of the sequent $\Gamma \Rightarrow \Delta$ and in the rest of this paper succedent contains at most one formula. A sequent $\Gamma \Rightarrow A$ can be read as "if all formulas in Γ hold then A holds." A sequent $\Gamma \Rightarrow$ can be read as "it cannot be the case that all formulas in Γ hold." We define $K\Gamma = \{KA | A \in \Gamma\}$.

Remark that the initial sequents, structural rules and logical rules are from the propositional part of Gentzen's **LJ**.

Definition 5 *Let $\Lambda \in \{$**IEL**, **IEL**$^-\}$. Let $\mathsf{G}(\Lambda)$ be one of systems of Table 2, $\mathsf{G}^-(\Lambda)$ be the systems without the (Cut) rule. A derivation $\mathscr{D}$ in $\mathsf{G}(\Lambda)$ (or $\mathsf{G}^-(\Lambda)$) is a finite tree generated by rules of $\mathsf{G}(\Lambda)$ (or $\mathsf{G}^-(\Lambda)$, respectively) from initial sequents. If there is a derivation in $\mathsf{G}(\Lambda)$(or $\mathsf{G}^-(\Lambda)$) that ends with a sequent $\Gamma \Rightarrow \Delta$, we say that $\Gamma \Rightarrow \Delta$ is derivable in $\mathsf{G}(\Lambda)$ (or $\mathsf{G}^-(\Lambda)$) which is denoted as $\mathsf{G}(\Lambda) \vdash \Gamma \Rightarrow \Delta$ (or $\mathsf{G}^-(\Lambda) \vdash \Gamma \Rightarrow \Delta$, respectively).*

Krupski and Yatmanov (2016) provided a sequent calculus of **IEL**. The sequent calculus is obtained from the propositional part of Gentzen's sequent calculus **LJ** plus the following two inference rules on the knowledge operator:

Cut-free and Analytic Sequent Calculus of Intuitionistic Epistemic Logic

Table 2 Sequent Calculi G(**IEL**$^-$) and G(**IEL**).

	Sequent Calculus G(**IEL**$^-$): Δ contains at most one formula below.
Initial Sequents	$A \Rightarrow A \qquad \bot \Rightarrow$
Structural Rules	$\dfrac{\Gamma \Rightarrow \Delta}{A,\Gamma \Rightarrow \Delta}$ (LW) $\quad \dfrac{\Gamma \Rightarrow}{\Gamma \Rightarrow C}$ (RW) $\quad \dfrac{A,A,\Gamma \Rightarrow \Delta}{A,\Gamma \Rightarrow \Delta}$ (LC)
	$\dfrac{\Gamma \Rightarrow A \quad A,\Gamma' \Rightarrow \Delta}{\Gamma,\Gamma' \Rightarrow \Delta}$ (Cut)
Logical Rules	$\dfrac{\Gamma \Rightarrow A_1 \quad \Gamma \Rightarrow A_2}{\Gamma \Rightarrow A_1 \wedge A_2}$ (R$\wedge$) $\quad \dfrac{A_i,\Gamma \Rightarrow \Delta}{A_1 \wedge A_2,\Gamma \Rightarrow \Delta}$ (L$\wedge$)
	$\dfrac{\Gamma \Rightarrow A_i}{\Gamma \Rightarrow A_1 \vee A_2}$ (R$\vee$) $\quad \dfrac{A_1,\Gamma \Rightarrow \Delta \quad A_2,\Gamma \Rightarrow \Delta}{A_1 \vee A_2,\Gamma \Rightarrow \Delta}$ (L$\vee$)
	$\dfrac{A,\Gamma \Rightarrow B}{\Gamma \Rightarrow A \supset B}$ (R$\supset$) $\quad \dfrac{\Gamma \Rightarrow A \quad B,\Gamma' \Rightarrow \Delta}{A \supset B,\Gamma,\Gamma' \Rightarrow \Delta}$ (L$\supset$)
Modal Rule	$\dfrac{\Gamma_1,\Gamma_2 \Rightarrow A}{\Gamma_1,K\Gamma_2 \Rightarrow KA}$ (K_{IEL^-})
	Sequent Calculus G(**IEL**): Δ contains at most one formula below. Replace (K_{IEL^-}) of G(**IEL**$^-$) with the following rule:
Modal Rules	$\dfrac{\Gamma_1,\Gamma_2 \Rightarrow \Delta}{\Gamma_1,K\Gamma_2 \Rightarrow K\Delta}$ (K_{IEL})

$$\dfrac{\Gamma_1,\Gamma_2 \Rightarrow A}{\Gamma_1,K\Gamma_2 \Rightarrow KA} \text{ (KI)} \qquad \dfrac{\Gamma \Rightarrow K\bot}{\Gamma \Rightarrow F.} \text{ (U)}$$

They established the cut-elimination theorem of the calculus. They also specify the computational complexity of the **IEL** as PSPACE-complete by a proof-theoretic method in terms of their sequent calculus. It is remarked, however, that this system does not enjoy the subformula property. That is, in the rule of (U), we have a formula $K\bot$ which might not be a subformula of a formula in the lower sequent of the rule (U).

Theorem 1 (The Equivalence of Hilbert and Gentzen Systems) *Let $\Lambda \in \{$**IEL**$^-,$ **IEL**$\}$, $\mathsf{H}(\Lambda) \vdash A$ if and only if $\mathsf{G}(\Lambda) \vdash \Rightarrow A$.*

Proof. We only show the case that the axiom (IR) is derivable in G(**IEL**).

$$\cfrac{\cfrac{\cfrac{\cfrac{\cfrac{A \Rightarrow A \quad \bot \Rightarrow}{A, \neg A \Rightarrow}(L \supset)}{KA, \neg A \Rightarrow}(K_{\text{IEL}})}{KA, \neg A \Rightarrow \bot}(RW)}{KA \Rightarrow \neg\neg A}(R \supset)}{\Rightarrow KA \supset \neg\neg A}(R \supset)$$

□

Definition 6 *Given a sequent $\Gamma \Rightarrow \Delta$, Γ_* denotes the conjunction of all formulas in Γ ($\Gamma_* \equiv \top$ if Γ is empty) and Δ^* denotes the unique formula in Δ ($\Delta^* \equiv \bot$ if Δ is empty). We say that a sequent $\Gamma \Rightarrow \Delta$ is valid in a class $\mathbb{M}$ of models (denoted by $\mathbb{M} \models \Gamma \Rightarrow \Delta$), if $\mathbb{M} \models \Gamma_* \supset \Delta^*$.*

Theorem 2 (Soundness) *Let $\Gamma \Rightarrow \Delta$ be a sequent.*

1. *If $\mathsf{G}(\mathbf{IEL}^-) \vdash \Gamma \Rightarrow \Delta$ then $\mathbb{M}_{\text{all}} \models \Gamma \Rightarrow \Delta$.*
2. *If $\mathsf{G}(\mathbf{IEL}) \vdash \Gamma \Rightarrow \Delta$ then $\mathbb{M}_{\text{ser}} \models \Gamma \Rightarrow \Delta$.*

We show that the cut elimination theorems hold for $\mathsf{G}(\mathbf{IEL}^-)$ and $\mathsf{G}(\mathbf{IEL})$, and we employ our proof-theoretic arguments from Ono and Komori (1985) and Kashima (2009). Recall that $\mathsf{G}^-(\mathbf{IEL}^-)$ and $\mathsf{G}^-(\mathbf{IEL})$ stand for the systems without (Cut) rules.

Theorem 3 (Cut-Elimination) *Let $\Gamma \Rightarrow \Delta$ be a sequent.*

1. *If $\mathsf{G}(\mathbf{IEL}^-) \vdash \Gamma \Rightarrow \Delta$ then $\mathsf{G}^-(\mathbf{IEL}^-) \vdash \Gamma \Rightarrow \Delta$.*
2. *If $\mathsf{G}(\mathbf{IEL}) \vdash \Gamma \Rightarrow \Delta$ then $\mathsf{G}^-(\mathbf{IEL}) \vdash \Gamma \Rightarrow \Delta$.*

Proof. We only show the case of $\mathsf{G}(\mathbf{IEL})$. For the sake of the contraction rules, we show the elimination of the extended form of (Cut) as $(Ecut)$ where $(Ecut)$ has the following form:

$$\cfrac{\cfrac{\vdots \mathscr{D}_1}{\Gamma \Rightarrow A} rule(D_1) \quad \cfrac{\vdots \mathscr{D}_2}{A^n, \Gamma' \Rightarrow \Delta} rule(D_2)}{\Gamma, \Gamma' \Rightarrow \Delta,}(Ecut)$$

where A^n ($n \geqslant 0$) means n-times repetition of the formula A and the formula A is called an *Cut formula* simply. It is noted that the ordinary (Cut) becomes a particular instance of $(Ecut)$. We show that, if an $(Ecut)$ only appears as the last step of a derivation $\mathscr{D}$, then there is a derivation in which no $(Ecut)$ appears and it ends with the same conclusion as $\mathscr{D}$.

This can be proved by double induction on the *complexity* (the number of logical connectives of the cut formulas) and the *weight*, i.e., the number of all the sequents in the derivation. We only show the case where both rules above the last application of $(Ecut)$ are rules of $K_{\mathbf{IEL}}$, i.e.,

$$\dfrac{\dfrac{\mathscr{D}_1}{\Gamma_1, \Gamma_2 \Rightarrow A} \quad (K_{\mathrm{IEL}})}{\Gamma_1, K\Gamma_2 \Rightarrow KA} \quad \dfrac{\dfrac{\mathscr{D}_2}{(A)^m, (KA)^n, \Gamma_3, \Gamma_4 \Rightarrow \Delta}}{(KA)^m, (KA)^n, \Gamma_3, K\Gamma_4 \Rightarrow K\Delta} \quad (K_{\mathrm{IEL}})}{\Gamma_1, K\Gamma_2, \Gamma_3, K\Gamma_4 \Rightarrow K\Delta} \quad (Ecut)$$

Then it suffices for us to transform this derivation into the following.

$$\dfrac{\mathscr{D}_1 \quad \dfrac{\dfrac{\mathscr{D}_1}{\Gamma_1, \Gamma_2 \Rightarrow A}}{\Gamma_1, K\Gamma_2 \Rightarrow KA} (K_{\mathrm{IEL}}) \quad \dfrac{\mathscr{D}_2}{(A)^m, (KA)^n, \Gamma_3, \Gamma_4 \Rightarrow \Delta}}{\dfrac{\dfrac{(A)^m, \Gamma_1, K\Gamma_2, \Gamma_3, \Gamma_4 \Rightarrow \Delta}{\Gamma_1, \Gamma_2, \Gamma_1, K\Gamma_2, \Gamma_3, \Gamma_4 \Rightarrow \Delta}}{\dfrac{\Gamma_1, K\Gamma_2, \Gamma_1, K\Gamma_2, \Gamma_3, K\Gamma_4 \Rightarrow K\Delta}{\Gamma_1, K\Gamma_2, \Gamma_3, K\Gamma_4 \Rightarrow K\Delta.} (LC^*)} (K_{\mathrm{IEL}})} (Ecut)}$$

where (LC^*) means finitely many applications of the rule (LC). □

Then, we obtain the following proof-theoretic results.

Corollary 1 (Subformula Property)   *Let* $\Lambda \in \{\mathbf{IEL}^-, \mathbf{IEL}\}$. *If* $\Gamma \Rightarrow \Delta$ *is derivable in* $\mathsf{G}(\Lambda)$ *by a derivation* $\mathscr{D}$*, then we can find a derivation* $\mathscr{D}'$ *ends with* $\Gamma \Rightarrow \Delta$ *such that* $A \in Sub(\Gamma \cup \Delta)$ *for any formula A from any sequent in* $\mathscr{D}'$*.*

Corollary 2 (Disjunction Property)   *Let* $\Lambda \in \{\mathbf{IEL}^-, \mathbf{IEL}\}$. *For any formulas A and B, if* $\Rightarrow A \vee B$ *is derivable in* $\mathsf{G}(\Lambda)$ *then either* $\Rightarrow A$ *or* $\Rightarrow B$ *is derivable in* $\mathsf{G}(\Lambda)$.

Next we prove Craig interpolation theorems by Maehara's method. In what follows, $\mathsf{V}(\Gamma)$ denotes the set of all propositional variables in the formulas in Γ. We say that $\langle \Gamma_1; \Gamma_2 \rangle$ is a *partition* of Γ if Γ is Γ_1, Γ_2.

Lemma 1   *Let* $\Lambda \in \{\mathbf{IEL}^-, \mathbf{IEL}\}$. *If* $\Gamma \Rightarrow \Delta$ *is derivable in* $\mathsf{G}(\Lambda)$*, then for any partition* $\langle \Gamma_1; \Gamma_2 \rangle$ *of* Γ *there exists a formula C* (*interpolant formula*) *such that both* $\Gamma_1 \Rightarrow C$ *and* $C, \Gamma_2 \Rightarrow \Delta$ *are also derivable in* $\mathsf{G}(\Lambda)$ *and* $\mathsf{V}(C) \subseteq \mathsf{V}(\Gamma_1) \cap \mathsf{V}(\Gamma_2, \Delta)$.

Proof. If $\mathsf{G}(\Lambda) \vdash \Gamma \Rightarrow \Delta$, then $\mathsf{G}^-(\Lambda) \vdash \Gamma \Rightarrow \Delta$ by Theorem 3. We prove by induction on a cut-free derivation of $\Gamma \Rightarrow \Delta$. We show the case where the last rule in the derivation is $(K_{\mathbf{IEL}})$:

$$\frac{\Gamma, \Gamma' \Rightarrow \Delta}{\Gamma, K\Gamma' \Rightarrow K\Delta}\ (K_{\mathbf{IEL}})$$

Let us consider the partition $\langle \Gamma_1, K\Gamma_1'; \Gamma_2, K\Gamma_2' \rangle$ of $\Gamma, K\Gamma'$. From the induction hypothesis, there exists a formula C such that $\Gamma_1, \Gamma_1' \Rightarrow C$ and $C, \Gamma_2, \Gamma_2' \Rightarrow \Delta$ are derivable and C satisfies the required conditions. By the following derivations:

$$\frac{\Gamma_1, \Gamma_1' \Rightarrow C}{\Gamma_1, K\Gamma_1' \Rightarrow KC}\ (K_{\mathbf{IEL}}) \qquad \frac{C, \Gamma_2, \Gamma_2' \Rightarrow \Delta}{KC, \Gamma_2, K\Gamma_2' \Rightarrow K\Delta}\ (K_{\mathbf{IEL}})$$

we can have an interpolant formula as KC where the condition of propositional variables holds trivially. □

Corollary 3 (Craig Interpolation Theorem) *Let* $\Lambda \in \{\mathbf{IEL}^-, \mathbf{IEL}\}$. *If* $\Rightarrow A \supset B$ *is derivable in* $\mathsf{G}(\Lambda)$, *then there exists a formula* C *such that* $\Rightarrow A \supset C$ *and* $\Rightarrow C \supset B$ *are derivable in* $\mathsf{G}(\Lambda)$ *and that* $\mathsf{V}(C) \subseteq \mathsf{V}(A) \cap \mathsf{V}(B)$.

These corollaries follow from the cut-elimination theorem since all inference rules except the cut rule satisfy the subformula property in our system.

4 Finite Model Property

This section establishes the finite model property of $\mathsf{G}(\mathbf{IEL}^-)$ and $\mathsf{G}(\mathbf{IEL})$. Let $\mathbb{M}_{\mathrm{all}}^{\mathrm{finite}}$ be the class of all finite models and $\mathbb{M}_{\mathrm{ser}}^{\mathrm{finite}}$ be the class of all serial finite models. In this section, even if Γ is a set of formulas, we regard $\Gamma \Rightarrow \Delta$ (where Δ contains at most one formula) as a sequent.

Definition 7 *Let* $\Lambda \in \{\mathbf{IEL}^-, \mathbf{IEL}\}$. *The sequent calculus* $\mathsf{G}(\Lambda)$ *enjoys the finite model property* with respect to a class $\mathbb{M}$ of finite frame *if* $\mathbb{M} \models \Gamma \Rightarrow \Delta$ *implies* $\mathsf{G}(\Lambda) \vdash \Gamma \Rightarrow \Delta$.

The finite model property for modal logic can be established, e.g., by a method from Takano (2018) in terms of sequent calculus. Moreover, the proof of semantic completeness of cut-free sequent calculus for intuitionistic logic can be carried in the setting of multi-succedent system (cf. Mints, 2000) or single-succedent system (cf. Hermant, 2005). We choose the combination of the second method by Hermant (2005), and the argument of Takano (2018).

Cut-free and Analytic Sequent Calculus of Intuitionistic Epistemic Logic

We establish the finite model property in an informative way, i.e., our proof also gives us an alternative semantic proof of the cut-elimination theorems.

Definition 8 *Let $\Lambda \in \{\mathbf{IEL}^-, \mathbf{IEL}\}$. Let Ω be a finite subformula closed set of formulas and $\Theta, \Pi \subseteq \Omega$ such that Π contains at most one formula. The set Θ is Π-saturated in $\mathsf{G}^-(\Lambda)$ for Ω if:*

1. $\mathsf{G}^-(\Lambda) \nvdash \Theta \Rightarrow \Pi$.

2. *for any formula $B \in \Omega$, either $\mathsf{G}^-(\Lambda) \vdash \Theta \cup \{B\} \Rightarrow \Pi$, or $B \in \Theta$.*

Lemma 2 (Lindenbaum Lemma) *Let $\Lambda \in \{\mathbf{IEL}^-, \mathbf{IEL}\}$. Let Ω be a finite subformula closed set of formulas and $\Gamma, \Pi \subseteq \Omega$ such that Π contains at most one formula. If $\mathsf{G}^-(\Lambda) \nvdash \Gamma \Rightarrow \Pi$, then there exists a finite set $\Theta \subseteq \Omega$ such that $\Gamma \subseteq \Theta$ and Θ is Π-saturated in $\mathsf{G}^-(\Lambda)$.*

Definition 9 *Given a set Γ of formulas, $K^-(\Gamma) := \{B | KB \in \Gamma\}$.*

Proposition 2 *Let $\Lambda \in \{\mathbf{IEL}^-, \mathbf{IEL}\}$. Let Ω be a finite subformula closed set of formulas, $\Theta, \Pi \subseteq \Omega$ and Π contains at most one formula. Suppose that Θ is Π-saturated in $\mathsf{G}^-(\Lambda)$ for Ω. Then the following holds.*

1. *if $B \wedge C \in \Theta$ then $B \in \Theta$ and $C \in \Theta$.*

2. *if $B \vee C \in \Theta$ then $B \in \Theta$ or $C \in \Theta$.*

3. *if $B \supset C \in \Theta$ then either $C \in \Theta$ either $\mathsf{G}^-(\Lambda) \nvdash \Theta \Rightarrow B$.*

4. *if $\mathsf{G}^-(\Lambda) \nvdash \Theta \Rightarrow B \wedge C$ then $\mathsf{G}^-(\Lambda) \nvdash \Theta \Rightarrow B$ or $\mathsf{G}^-(\Lambda) \nvdash \Theta \Rightarrow C$*

5. *if $\mathsf{G}^-(\Lambda) \nvdash \Theta \Rightarrow B \vee C$ then $\mathsf{G}^-(\Lambda) \nvdash \Theta \Rightarrow B$ and $\mathsf{G}^-(\Lambda) \nvdash \Theta \Rightarrow C$*

6. *if $\mathsf{G}^-(\Lambda) \nvdash \Theta \Rightarrow B \supset C$ then $\mathsf{G}^-(\Lambda) \nvdash \Theta, B \Rightarrow C$*

7. *if $\mathsf{G}^-(\Lambda) \nvdash \Theta \Rightarrow KB$ then $\mathsf{G}^-(\Lambda) \nvdash \Theta \cup K^-(\Theta) \Rightarrow B$.*

Proof. We prove item 7 alone. Suppose $\mathsf{G}^-(\Lambda) \nvdash \Theta \Rightarrow KB$. Assume for contradiction that $\mathsf{G}^-(\Lambda) \vdash \Theta \cup K^-(\Theta) \Rightarrow B$. Then we obtain $\mathsf{G}^-(\Lambda) \vdash \Theta, KK^-(\Theta) \Rightarrow KB$. Moreover, we apply the contraction rules finitely many times to obtain $\mathsf{G}^-(\Lambda) \vdash \Theta \Rightarrow KB$. A contradiction with the assumption $\mathsf{G}^-(\Lambda) \nvdash \Theta \Rightarrow KB$. □

Definition 10 *Let $\Lambda \in \{\mathbf{IEL}^-, \mathbf{IEL}\}$. Given a finite subformula closed set Ω of formulas, the Ω-model $M_\Omega = (W, \leq, R, V)$ for $\mathsf{G}^-(\Lambda)$ is defined as follows:*

- $W := \{\Theta \subseteq \Omega : \Theta \text{ is a } \Pi\text{-saturated for } \Omega \text{ in } \mathsf{G}^-(\Lambda), \text{ for some } \Pi \subseteq \Omega \text{ such that } \Pi \text{ contains at most one formula }\}$,

- $\Theta_1 \leq \Theta_2$ iff $\Theta_1 \subseteq \Theta_2$,

- $\Theta_1 R \Theta_2$ iff $K^-(\Theta_1) \cup \Theta_1 \subseteq \Theta_2$.

- $\Theta \in V(p)$ iff $p \in \Theta$.

Proposition 3 *Let Ω be a finite subformula closed set.*

1. *The Ω-model M_Ω for $\mathsf{G}(\mathbf{IEL}^-)$ is in $\mathbb{M}_{\text{all}}^{\text{finite}}$.*

2. *The Ω-model M_Ω for $\mathsf{G}(\mathbf{IEL})$ is in $\mathbb{M}_{\text{ser}}^{\text{finite}}$.*

Proof. We show the seriality of the latter statement alone. Fix any $\Theta \in W$ of M_Ω. Then we have $\mathsf{G}^-(\mathbf{IEL}) \nvdash \Theta \Rightarrow \Delta$ hence $\mathsf{G}^-(\mathbf{IEL}) \nvdash \Theta \cup K^-(\Theta) \Rightarrow$ by ($K_{\mathbf{IEL}}$). Now we can construct an $\emptyset$-saturated Π in $\mathsf{G}^-(\mathbf{IEL})$ for Ω. Therefore $\Theta \cup K^-(\Theta) \subseteq \Pi$ hence $\Theta R \Pi$. □

Lemma 3 (Truth Lemma) *Let $\Lambda \in \{\mathbf{IEL}^-, \mathbf{IEL}\}$, Ω be a finite subformula closed set of formulas and $\Theta \in W$ of M_Ω. For all $A \in \Omega$,*

1. *if $A \in \Theta$ then $M_\Omega, \Theta \models A$.*

2. *if $\mathsf{G}^-(\Lambda) \nvdash \Theta \Rightarrow A$ then $M_\Omega, \Theta \nvDash A$.*

Proof. Induction on the complexity of A. Let $\Lambda = \mathbf{IEL}$. We only show the case where A is in the form of KC. First we establish term 1. Suppose $KC \in \Theta$, we show that $M_\Omega, \Theta \models KC$. Fix any Δ, suppose $\Theta R \Delta$, we show $M_\Omega, \Delta \models C$. From $\Theta R \Delta$ we have $K^-(\Theta) \subseteq \Delta$. Since $C \in K^-(\Theta)$, we have $C \in \Delta$. Then we have $M_\Omega, \Delta \models C$ by induction hypothesis.

For term 2, suppose $\mathsf{G}^-(\mathbf{IEL}) \nvdash \Theta \Rightarrow KC$, we show $M_\Omega, \Theta \nvDash KC$. By Proposition 2, we have $\mathsf{G}^-(\mathbf{IEL}) \nvdash \Theta \cup K^-(\Theta) \Rightarrow C$. Then by Lindenbaum Lemma, we have a C-saturated Δ, such that $\Theta \cup K^-(\Theta) \subseteq \Delta$ and $\mathsf{G}^-(\mathbf{IEL}) \nvdash \Delta \Rightarrow C$. From $\Theta \cup K^-(\Theta) \subseteq \Delta$ we have $\Theta R \Delta$, from induction hypothesis we have $M_\Omega, \Delta \nvDash C$. Then $M_\Omega, \Theta \nvDash KC$ holds. □

Theorem 4 *Let Δ contain at most one formula,*

1. *if $\mathbb{M}_{\text{all}}^{\text{finite}} \models \Gamma \Rightarrow \Delta$ then $\mathsf{G}^-(\mathbf{IEL}^-) \vdash \Gamma \Rightarrow \Delta$;*

2. *if $\mathbb{M}_{\text{ser}}^{\text{finite}} \models \Gamma \Rightarrow \Delta$ then $\mathsf{G}^-(\mathbf{IEL}) \vdash \Gamma \Rightarrow \Delta$.*

Proof. We only show the case of **IEL**. Suppose $\mathsf{G}^-(\mathbf{IEL}) \nvdash \Gamma \Rightarrow \Delta$, put $\Omega = Sub(\Gamma, \Delta)$ which is a finite subformula closed set. We construct a set of formulas $\Theta \subseteq \Omega$ such that, $\Gamma \subseteq \Theta$ and Θ is Δ-saturated for Ω in $\mathsf{G}^-(\mathbf{IEL})$. Note that $\Theta \in W$ in M_Ω. By Truth Lemma, $M_\Omega \nvDash \Gamma \Rightarrow \Delta$. From $M_\Omega \in \mathbb{M}_{\text{ser}}^{\text{finite}}$, $\mathbb{M}_{\text{ser}}^{\text{finite}} \nvDash \Gamma \Rightarrow \Delta$. □

Corollary 4 *Let $\Lambda \in \{\mathbf{IEL}, \mathbf{IEL}^-\}$. The following are all equivalent.*

1. $\mathbb{M}_\Lambda \vDash A$;
2. $\mathbb{M}_\Lambda^{\text{finite}} \vDash A$;
3. $\mathsf{G}^-(\Lambda) \vdash \Rightarrow A$;
4. $\mathsf{G}(\Lambda) \vdash \Rightarrow A$;
5. $\mathsf{H}(\Lambda) \vdash A$,

where $\mathbb{M}_{\mathbf{IEL}^-}^{(\text{finite})} := \mathbb{M}_{\text{all}}^{(\text{finite})}$ *and* $\mathbb{M}_{\mathbf{IEL}}^{(\text{finite})} := \mathbb{M}_{\text{ser}}^{(\text{finite})}$.

Proof. The direction from item 1 to item 2 is trivial. The direction from item 2 to item 3 can be obtained from Theorem 4. The direction from item 3 to item 4 is trivial. The direction from item 4 to item 5 can be obtained from the equivalence of the Hilbert systems and the sequent calculi (Theorem 1). The direction from item 5 to item 1 can be obtained from the Fact 1. □

In particular, we can also prove the cut elimination theorems semantically by the soundness result of the sequent calculi (Theorem 2) and Theorem 4.

References

Artemov, S., & Protopopescu, T. (2016). Intuitionistic epistemic logic. *The Review of Symbolic Logic, 9*, 266–298.
DeVidi, D., & Solomon, G. (2001). Knowability and intuitionistic logic. *Philosophia, 28*, 319–334.
Dummett, M. (1978). Truth and other enigmas. In *New Essay on the Knowability Paradox* (pp. 145–165). Duckworth.
Dummett, M. (2009). Fitch's paradox of knowability. In J. Salerno (Ed.), *New Essay on the Knowability Paradox* (pp. 51–52). Oxford University.
Fitch, F. (1963). A logical analysis of some value concepts. *Journal of Symbolic Logic, 28*(2), 135–142.
Hermant, O. (2005). Semantic cut elimination in the intuitionistic sequent calculus. In P. Urzyczyn (Ed.), *Typed Lambda-Calculi and Applications* (Vol. 3461, pp. 221–233). Nara, Japan: Springer-Verlag.

Kashima, R. (2009). *Mathematical logic*. Asakura Publishing Co. Ltd (in Japanese).

Krupski, V. N., & Yatmanov, A. (2016). Sequent calculus for intuitionistic epistemic logic IEL. In *International Symposium on Logical Foundations of Computer Science* (pp. 187–201). Springer.

Mints, G. (2000). *A Short Introduction to Intuitionistic Logic*. Springer US.

Ono, H., & Komori, Y. (1985). Logics without the contraction rule. *The Journal of Symbolic Logic*, *50*(1), 169–201.

Proietti, C. (2012). Intuitionistic epistemic logic, Kripke models and Fitch's paradox. *Journal of Philosophical Logic*, *41*, 877–900.

Su, Y., & Sano, K. (2020). Logics for knowability paradox with a non-normal possibility operator. In F. Liu, H. Ono, & J. Yu (Eds.), *Knowledge, Proof and Dynamics* (pp. 51–72). Singapore: Springer Singapore.

Sundholm, G. (2014). Constructive recursive functions, Church's thesis, and Brouwer's theory of the creating subject: Afterthoughts on a parisian joint session. In J. Dubucs & M. Bourdeau (Eds.), *Constructivity and Computability in Historical and Philosophical Perspective* (pp. 1–35). Dordrecht: Springer Netherlands.

Takano, M. (2018). A semantical analysis of cut-free calculi for modal logics. *Reports on Mathematical Logic*, *53*, 43–65.

van Dalen, D. (2013). *Logic and structure* (5th ed.). Springer-Verlag.

Williamson, T. (1992). On intuitionistic modal epistemic logic. *Journal of Philosophical Logic*, *21*, 63–89.

Youan Su
Hokkaido University, Graduate School of Letters
Japan
E-mail: `ariyasu613@gmail.com`

Katsuhiko Sano
Hokkaido University, Faculty of Humanities and Human Sciences
Japan
E-mail: `v-sano@let.hokudai.ac.jp`

Do We Need Recursion?

VÍTĚZSLAV ŠVEJDAR

Abstract: The operation of primitive recursion, and recursion in a more general sense, is undoubtedly a useful tool. However, we will explain that in two situation where we work with it, namely in the definition of partial recursive functions and in logic when defining the basic syntactic notions, its use can be avoided. We will also explain why one would want to do so.

Keywords: recursion, bounded formulas, formalized syntax

1 What is recursion, where do we meet it?

Recursion in a narrow sense is an operation with multivariable functions whose arguments are natural numbers. Let $\underline{z}$ denote the k-tuple $z_1, \ldots, z_k$ (where the number k does not have to be indicated), and let $\widehat{f}(x, \underline{z})$ denote the code $\langle f(0, \underline{z}), \ldots, f(x-1, \underline{z}) \rangle$ of the sequence $f(0, \underline{z}), \ldots, f(x-1, \underline{z})$ under some suitable coding of finite sequences. If (1), (2) or (3) holds for any choice of arguments:

$$f(0) = a, \qquad f(x+1) = h(f(x), x), \tag{1}$$

$$f(0, \underline{z}) = g(\underline{z}), \qquad f(x+1, \underline{z}) = h(f(x, \underline{z}), x, \underline{z}), \tag{2}$$

$$f(x, \underline{z}) = h(\widehat{f}(x, \underline{z}), \underline{z}), \tag{3}$$

then in the cases (1) and (2) we say that f is derived from a number a and a function h or from two functions g and h by *primitive recursion*, while in the case (3) we say that f is derived from h by *course-of-values* recursion. All functions in (1)–(3) may be partial, i.e., undefined for some arguments. Nevertheless, if h is (both g and h are) total (everywhere defined), then the derived function f must be total. It does not matter that x does not appear in (3) as a variable of h because it can be determined as the length of the sequence encoded in the number $\langle f(0, \underline{z}), \ldots, f(x-1, \underline{z}) \rangle$. Course-of-values recursion seems more powerful, but it can be simulated by (derived from) primitive recursion of the form (1) and (2). Therefore, it can be considered a variant of primitive recursion.

Vítězslav Švejdar

Primitive recursion appears in very basic definitions: a function is *partial recursive* if it can be derived from the initial functions using composition, minimization and primitive recursion; it is *primitive recursive* if it can be derived from the same initial functions using composition and primitive recursion only. Since *RE* sets (recursively enumerable sets) are usually defined as the domains of partial recursive functions, the operation of primitive recursion is in fact part of the definition of the arithmetical hierarchy. Primitive recursion is also a useful tool in some proofs. For example, if g is a recursive function with an infinite range, then the equation $f(x) = g(\mu v(g(v) \notin \{f(0),\ldots,f(x-1)\}))$ where $\mu v(..)$ denotes the minimization operation (that is, the search for the first v satisfying the condition in the parentheses) is a derivation by course-of-values recursion of a one-to-one function with the same range. This argument in fact shows that any infinite *RE* set is the range of some one-to-one recursive function.

Recursion in a broader sense is used in *programming languages*: a procedure can be written so that it processes its parameter by calling itself, perhaps several times, with parameters that are simpler in some sense. The parameters do not have to be natural numbers. Also in *logic* we have several definitions that are described as recursive. One example is this: an expression is a term of a language L if it is a variable, or if it is a constant, or if it has the form $F(t_1,\ldots,t_n)$ where $F \in L$ is an n-ary function symbol and $t_1,\ldots,t_n$ are terms. This and other syntactic definitions deal with strings rather that numbers, and they are thus examples of recursion in a broader sense. However, if syntactic objects are identified with natural numbers via some coding of syntax, all these definitions appear to be applications of course-of-values recursion. Then, the set of all terms, the set of all formulas, etc. are primitive recursive sets.

When dealing with metamathematics of Peano arithmetic PA and with incompleteness phenomena, one may need (in fact, does need) arithmetic formulas that define *RE* sets. When dealing with Gödel's second incompleteness theorem, one (of course) needs logical syntax formalized inside Peano arithmetic. In both situations primitive recursion poses a problem because in the arithmetic language (consisting of the binary operation symbols $+$ and $\cdot$, the order symbols $<$ and $\leq$, the constant 0 and the successor function S) there is nothing that would directly correspond to it. One can use existential quantifiers and describe a function that is derived by composition (from functions that can be described), and one can use the least number principle to describe a function that is derived by minimization. However, the dynamic nature of primitive recursion is problematic for the language of

Do We Need Recursion?

a formal theory where we primarily have static descriptions. S. Feferman in his paper (Feferman, 1960), which for decades was the most important source of information about Gödel's theorems and about interpretability, introduces the notion of PR-formulas. The purpose of this notion is to have a class of formulas that define and describe (formalize) exactly the primitive recursive conditions. Nevertheless, PR-formulas are more an *ad hoc* technical solution than wisely chosen notion that can be further studied. The formalization of syntax itself, and the complexity of the corresponding formulas, cannot be easily learnt from that paper.

While a brief answer to the question whether we need recursion is yes, we will explain that its use in the basic definitions and when formalizing syntax can be replaced by the use of Δ_0 conditions. The class of all Δ_0 conditions is somewhat smaller than the class of all primitive recursive conditions, but it is still quite expressive. It does have a counterpart in the formalized arithmetic, namely the class of Δ_0-formulas. Defining partial recursive functions without recursion may sound paradoxical, and showing that the defined concept remains unchanged does require some effort, but the fact that all *RE* sets are definable in the structure $\mathbb{N}$ of natural numbers then comes practically for free.

All claims and results of this paper are given in more detail in the book (Švejdar, 2020). Practically all ideas concerning formalized syntax are due to Pavel Pudlák: they are either contained in (Hájek & Pudlák, 1993) or were communicated otherwise. We just add some computations and offer detailed implementation. Defining partial recursive functions without primitive recursion is described in (Odifreddi, 1989).

2 Bounded conditions and bounded formulas

(Multivariable) polynomials in the domain $\mathbb{N}$ of natural numbers are functions like $[x, y] \mapsto 2x^2 + 3xy + 1$, obtained from variables and constants by (repeated use of) addition and multiplication. A *bounded condition* (or Δ_0 *condition*) is a condition obtained from equalities of polynomials using Boolean operations and the *bounded quantifiers* $\forall v \leq f(\underline{x})$, $\exists v \leq f(\underline{x})$, $\forall v < f(\underline{x})$ and $\exists v < f(\underline{x})$ where f is a polynomial not dependent on v. Bounded quantifiers have the obvious meaning: $\forall v \leq f(\underline{x}) A(v, \underline{x})$ is a shorthand for $\forall v(v \leq f(\underline{x}) \Rightarrow A(v, \underline{x}))$ etc. Notice that they interact the expected way with negation: $\neg \forall v \leq f(\underline{x}) A(v, \underline{x})$ and $\exists v \leq f(\underline{x}) \neg A(v, \underline{x})$ are equivalent, and similar equivalences hold for the remaining cases. Since one

Vítězslav Švejdar

can see a for-loop behind each bounded quantifier in a Δ_0 condition $A(\underline{x})$, it is easy to imagine a program (written in any reasonable programming language) that decides A.

An example of a Δ_0 condition is the divisibility relation: x is a divisor of y (written as $x \mid y$) if $\exists v {\leq} y\, (v \cdot x = y)$. Also being a prime is a Δ_0 condition because x is a prime if $x > 1$ & $\forall v {<} x\, (v \mid x \Rightarrow v = 1)$. Euclidean division, i.e., the two functions Mod and Div that yield the remainder and the quotient of dividing x by z, have Δ_0 graphs. In the case of the function Div the graph is $\{\, [y, x, z]\,;\, \exists r {<} x\, (x = y \cdot z + r) \lor (z = 0\ \&\ y = 0)\,\}$, where the purpose of the clause $(z = 0\ \&\ y = 0)$ is to have a total function, which yields some (unimportant) output even when the divisor z is zero. The fact that $y = \text{Mod}(x, z)$ is a Δ_0 condition is proved similarly.

Other examples are the graph $\{\,[y, x]\,;\, y = 2^x\,\}$ of the function $x \mapsto 2^x$ and the range Pwr of this function, i.e., the set $\{1, 2, 4, 8, \dots\}$ of all *powers of two*. Since Pwr $= \{\, y\,;\, \exists x {<} y\, (y = 2^x)\,\}$, the set (property) Pwr is obtained from the condition $y = 2^x$ using a bounded quantifier, and thus if this condition is Δ_0, then also Pwr $\in \Delta_0$. This is a sound but misleading observation. The problem about the condition $y = 2^x$ is that it is an equality, but not an equality of polynomials. Thus the fact that it is Δ_0 is a nontrivial result. For a proof see Bennet (1962) or (Pudlák, 1983). Below we will sketch a yet another proof. A direct proof that Pwr $\in \Delta_0$ is here: x is a power of two if and only if $\forall v {\leq} x\, (v \mid x \to v = 1 \lor 2 \mid v)$.

If r is a power of two, then $\text{Div}(\text{Mod}(u, 2r), r)$ yields the bit in the binary expansion of u that corresponds to the power r. Therefore, using Euclidean division and employing powers of two as pointers, we can speak (in the Δ_0 speech) about binary expansions of numbers. We can, for example, say that the bits corresponding to powers $r_1 < r_2$ are positive while all bits between them are negative. We cannot (yet) say that d is the x-th digit (because that would involve saying that $r = 2^x$). For our proof-sketch that not only $y = 2^x$, but also $y = z^x$ are Δ_0 conditions, consider the following "data structure" consisting of three numbers u, v and w:

$$
\begin{array}{c}
\overbrace{}^{x} \\
\underbrace{}_{y}\quad \underbrace{1101\,0000000110\,00011\,01}\quad u \\
\underline{11000010100111101001\overline{1}\,\,1011011001\,11011\,11\quad v} \\
1\,00000000000000000001\,0000000001\,00001\,01 \quad w \\
\uparrow \uparrow \\
r_2 r_1
\end{array}
\qquad (4)
$$

The number w acts as a ruler: the positive bits in its expansion are markers that divide the numbers u and v into items. The separation into items is also indicated by little gaps between digits. The items in u (looking from the right) are the numbers 1, 3, 6 and 13; their binary representations are 1, 11, 110, 1101. The items in v are 3, 27 (whose expansion is 11011), etc. Let $\text{ExpW}(y, x, z, u, v, w)$ be a shorthand for the condition describing the data structure of which (4) is an instance. The condition says "the lowest item in u is 1; the lowest item in v is z and $z > 1$; whenever an item in u is t and the corresponding item in v is e, then the next items in u and v are either $2t$ and e^2 or $2t+1$ and ze^2; the last (highest) items in u and v are x and y". The condition $\text{ExpW}(y, x, z, u, v, w)$ is Δ_0. If it holds, then u, v and w witness (for $x \neq 0$ and $z > 1$) that $y = z^x$. The condition:

$$\exists u \exists v \exists w \, \text{ExpW}(y, x, z, u, v, w) \,\vee \\ \vee \, (x = 0 \ \& \ y = 1) \vee (x \neq 0 \ \& \ z < 2 \ \& \ y = z) \quad (5)$$

deals also with the marginal cases where $x = 0$ or $z \leq 1$, and it defines the graph of the function $[x, z] \mapsto z^x$. It can be verified that if u, v and w are such that $\text{ExpW}(y, x, z, u, v, w)$, then they do not exceed y^3. This means that the three quantifiers in (5) can be written as $\exists u {\leq} y^3$, $\exists v {\leq} y^3$ and $\exists w {\leq} y^3$. Thus indeed, the condition $y = z^x$ is Δ_0.

The above proof is different from that in (Pudlák, 1983). But its main idea (to achieve an efficient data structure by using the recursive conditions $z^{2t} = (z^t)^2$ and $z^{2t+1} = z(z^t)^2$ rather than the condition $z^{t+1} = z(z^t)$) is also due to Pavel Pudlák.

A yet another useful function is the function $x \mapsto \text{NPB}(x)$ that yields the number of positive bits in the binary expansion of a number x. To show that it has a Δ_0 graph, we again design data structure that witnesses that $y = \text{NPB}(x)$. The data structure now consists of a single number w whose binary expansion is seen as a concatenation of (binary expansions of) numbers $S(w, r, i, j)$ that for $0 \leq i < r$ and $0 \leq j < 2^{r-i-1}$ satisfy $S(w, r, i+1, j) = S(w, r, i, 2j) + S(w, r, i, 2j+1)$, the number r is a power of two not smaller than the total number of bits in x, and $S(w, r, 0, j)$ for each $j < 2^r$ is the j-th digit of x (looking from the right). We call w *summation tree* for x. The numbers $S(w, r, i, j)$ for $i \leq r$ and $j < 2^{r-i}$ can be seen as labels of nodes in a binary tree: there are 2^r leaves labeled by the bits in x, each non-leaf is labeled by the sum of the labels of the two children, and the root is labeled by $S(w, r, r, 0)$, which is the result of the whole computation. For example, one can check that the binary expansion of the number $x = 24\,308\,687$ consists of 25 bits. Then the least possible r

Vítězslav Švejdar

is 5 and the summation tree w corresponding to x and this r is:

$$\overbrace{0000000\,1011100\,1011101\,01111\,001111}^{x} \quad 0$$
$$00\,00\,00\,01\,01\,10\,00\,01\,10\,01\,01\,10\,10\,00\,10\,10 \quad 1$$
$$000\,001\,011\,001\,011\,011\,010\,100 \quad 2$$
$$0001\,0100\,0110\,0110 \quad 3$$
$$00101\,01100 \quad 4$$
$$10001 \quad 5$$

where w is split across six lines (with more significant bits in lower lines). Again, little gaps between digits (together with line breaks) indicate the items $S(w, r, i, j)$ of w. There are $i + 1$ digits reserved for $S(w, r, i, j)$; the leading zeros in lines $i < r$ are (must be) given, but the leading zero in line 5 is not given because it would be a leading zero in the entire number w. The binary expansion of $S(w, r, r, 0)$ is 10001; indeed, there are 17 positive bits in x. For a full proof that "w is a summation tree for x" is a Δ_0 condition see (Švejdar, 2020). It is clear already now that the proof uses the fact that $y = 2^x$ is a Δ_0 condition. It can be proved that for every x there exists a summation tree w such that $w < x^8$, from which it follows that $y = \mathrm{NPB}(x)$ is a Δ_0 condition. The key idea of the proof of this fact is natural: if the positive bits in x are counted not one by one, but by dividing (repeatedly) the binary expansion of x into halves, then the data structure corresponding to the computation is efficient in the sense that it is bounded by a polynomial in x. It should be emphasized that, while we use some ideas from computational complexity and the word 'polynomial' appears several times in this paper, the class of all Δ_0 conditions is different from the classes *NP* and *P* studied in computational complexity.

Knowing that Δ_0 conditions express many useful properties and relations, we are ready to explain their relationship to computability theory and to arithmetic. As already mentioned, partial recursive functions are usually defined as the functions that can be derived from the initial functions by primitive recursion, composition and minimization where the initial functions are $x \mapsto 0$, $x \mapsto x + 1$ and $[x_1, .., x_k] \mapsto x_j$ for $1 \leq j \leq k$. It can be shown that if the list of the initial functions is extended by adding addition and multiplication and the function e where $\mathrm{e}(x, y) = 1$ if $x = y$ and $\mathrm{e}(x, y) = 0$ otherwise, then every partial recursive function can be derived without using primitive recursion. This fact is shown in (Odifreddi, 1989)

Do We Need Recursion?

(with citations to Gödel and Kleene). Working with this definition, one can show that the graph (and thus also the domain and the range) of every partial recursive function is a projection of (i.e., a condition obtained by existential quantification from) some Δ_0 relation. On the other hand, it is easy to prove (not employing primitive recursion) that every projection of a Δ_0 relation is the domain of some partial recursive function. From these considerations it follows that the basic notions of computability theory, partial recursive functions and *RE* sets, can be defined without primitive recursion (and actually also without using the functions $[x, z] \mapsto z^x$ and NPB; these are not needed until the next section).

From now on we deal with connections to formal arithmetic. We reserve the letters x, y, etc. for variables in (arithmetic) formulas. Out of many structures, we in fact only need the *standard model* $\mathbb{N} = \langle N, +, \cdot, 0, S, \leq, < \rangle$ of Peano arithmetic PA. We use n, k, etc. to denote its elements. We keep in mind that PA is incomplete and has many other models. For $n \in N$, the n-th *numeral* $\overline{n}$ is the closed term $S(S \ldots (0) \ldots)$ containing exactly n occurrences of the successor symbol S. Recall that a formula $\varphi(\underline{x})$ *defines* a set $A \subseteq N^k$ in $\mathbb{N}$ if the equivalence $A(n_1, \ldots, n_k) \Leftrightarrow \mathbb{N} \models \varphi(\overline{n_1}, \ldots, \overline{n_k})$ holds for every k-tuple $[n_1, \ldots, n_k]$. We introduce *bounded quantifiers*: these are quantifiers of the form $\forall v {\leq} t(\underline{x})$, $\exists v {\leq} t(\underline{x})$, $\forall v {<} t(\underline{x})$ and $\exists v {<} t(\underline{x})$ where $t(\underline{x})$ is an arithmetic term not containing v. We define *bounded formulas*, or Δ_0-*formulas*, as formulas in which all quantifiers are bounded. Notice that we use 'Δ' at the formal level and the italic 'Δ' at the meta level. The two notions correspond; the sets definable by Δ_0-formulas are exactly the Δ_0 relations, which can be written as follows: $\Delta_0^{\mathbb{N}} = \Delta_0$. This is so because terms in the arithmetic language correspond to multivariable polynomials.

Let $\mathsf{Pwr}(x)$ be the formula $\forall v {\leq} x \, (v \mid x \rightarrow v = \overline{1} \vee \overline{2} \mid v)$ where $v \mid x$ is the formula $\exists u {\leq} v \, (u \cdot v = x)$. The formula $\mathsf{Pwr}(x)$ is bounded and defines the set Pwr of all power of two. All our remaining considerations about Δ_0 conditions can be reproduced at the formal level. Thus we can introduce a bounded formula saying that the number y is the number of positive bits in x and write it as $y = \mathsf{NPB}(x)$, and we can introduce a bounded formula $y = z^x$ that describes exponentiation. We also have bounded formulas $y = \mathsf{Mod}(x, z)$ and $y = \mathsf{Div}(x, z)$ that describe Euclidean division. Notice that we use the single arrow $\rightarrow$ for implication as a symbol in formulas, while the double arrow $\Rightarrow$ was used at the meta level as a shorthand for implication in our speech. We also use the sans-serif font in the shorthands for formulas and also in informal readings of formulas (one should always imagine a formula behind the sans-serif font). We do not systematically invent double symbols: we write the connectives $\&$, $\vee$ and $\neg$, the divisibility symbol $|$ and equations like $y = z^x$ the same way at both the formal and the meta level. We do not add new function symbols to the arithmetic language; equations like $y = \mathsf{Mod}(x, z)$ are shorthands for formulas, not equalities of terms.

As already noted, the formula $\mathsf{Pwr}(x)$ defines the set of all powers of two, which means that any n is in Pwr if and only if $\mathbb{N} \models \mathsf{Pwr}(\overline{n})$. The same holds for the remaining formulas. Thus, for example, $\mathbb{N} \models \overline{m} = \mathsf{NPB}(\overline{n})$ if and only if $m = \mathsf{NPB}(n)$, i.e., if and only if the number of positive bits in the binary expansion of n is exactly m. However, our Δ_0-formulas *not only define* the sets that they are supposed to define. They *formalize* the corresponding notions in Peano arithmetic. That is, if they are accepted as definitions of the notions inside PA, then PA can *prove* the expected properties of the notions. "Expected properties" is a vague concept, but in most cases it is clear what to expect. PA can prove that every divisor of a power of two is a power of two and that the product of any two powers of two is again a power of two. As to the exponential function, $\mathsf{PA} \vdash \forall x \forall y \forall z (z^{x+y} = z^x \cdot z^y)$ and also $\mathsf{PA} \vdash \forall x \forall y \forall z ((z^y)^x = z^{y \cdot x})$. The function NPB has the following summation property: if s is a power of two greater than x, then $\mathsf{NPB}(y \cdot s + x) = \mathsf{NPB}(y) + \mathsf{NPB}(x)$. It is important to realize that provability (in a theory in general, and in PA in particular) is the same as validity in *all models*. Thus it is good to think of powers of two, the arguments of the exponential function etc. as nonstandard elements of some model of PA. If $\mathcal{M}$ is such a model, it contains nonstandard powers of two, but they behave as expected. Binary expansions can have infinite (nonstandard) length but the positive and negative bits in them can be counted, and $z^{x+y} = z^x \cdot z^y$ holds whether x, y and z are standard or not.

3 Arithmetization of syntax

In this section we deal with syntactic notions formalized in PA. That is, we start with variables and terms and proceed to proofs and provability. We assume that the theory being formalized is PA itself. A straightforward modification of our considerations would make it possible to work in PA with provability in some other theory, say in the Zermelo-Fraenkel set theory ZF.

Since in PA we have numbers and nothing else, we have to specify coding of syntactic objects with numbers. We opt for a coding method that we find natural: characters are identified with their numerical codes, the nu-

Do We Need Recursion?

merical codes are specified by the *code table* whose size is b, characters in a string w are the digits in the b-ary expansion of w. In more detail, we assume that b = 128 and that the code table is a modified ASCII table. The codes 32–126 are the same as in the ASCII table, the slots 1–31 and 127 (in which the ASCII table has invisible characters) contain characters that are needed in logic but do not occur in the ASCII table (the quantifiers, the set epsilon, ...), no code is zero. For example, if w is the number $83 \cdot b^6 + 40 \cdot b^5 + 83 \cdot b^4 + 40 \cdot b^3 + 48 \cdot b^2 + 41 \cdot b + 41$, then w is the code of the string S(S(0)) because 83, 40, 48 and 41 are the codes (in both the code table and in the ASCII table) assigned to the characters S, (, 0 and). We write the characters themselves, typeset in the typewriter font, instead of their numerical codes, and we identify strings with their numerical codes. Thus if w is still the same number, we can write w = S(S(0)). We omit some of the bars that indicate numerals if there is no danger of misunderstanding.

A number is a *string* if no digit in its b-ary expansion is zero. The *length* Lh(w) of a string w is the least y such that $w < b^y$. The x-th symbol $(w)_x$ of a string w (looking from the left and counting from zero) is the number Div(Mod(w, $b^{\mathsf{Lh}(w)-x}$), $b^{\mathsf{Lh}(w)-x-1}$) for $x <$ Lh(w), and it is 0 otherwise. Thus for example, the number 0 is the empty string and its length is 0. The number 0 (notice the typewriter font) is a string having length 1, and we have $(0)_0 = 0$. We also have Lh(S(S(0))) = 7 and $(\mathtt{S(S(0))})_1 = (\mathtt{S(S(0))})_3 = 40$ because 40 is the numerical code of the character (. Being a string is obviously a Δ_0-formula. Since $y = z^x$ is a Δ_0-formula, also $y =$ Lh(w) and $y = (w)_x$ are Δ_0-formulas. The *concatenation* $w_1 * w_2$ of w_1 and w_2 is the number $w_1 \cdot b^{\mathsf{Lh}(w_2)} + w_2$. We omit the symbol $*$ (and write just $w_1 w_2$) if it is clear that we deal with strings.

We need infinitely many variables when writing formulas, and we can assume that there are only countably many of them and that they are indexed by natural numbers. Therefore, we define that a *variable* is a nonempty string consisting of the letter v followed by a binary expansion of a number. A binary expansion of a number is a string s consisting of the digits 0 and 1 such that the leftmost digit of s can be 0 only if Lh(s) = 1. Let Var(x) be the formula the number x is a variable. Clearly, Var(x) is a Δ_0-formula. It does not matter (it does not complicate syntax analysis) that the character 0 plays a double role: it appears in indices of variables, and it is also a constant in the arithmetic language. For example, the string version of the formula $\neg \exists v_2 (v_2 = 0)$ is ¬∃v10(v10=0).

Let NOcc(u, w) be the *number of occurrences* of the character u in a string w. We again want the function to be total, and thus we assume that

Vítězslav Švejdar

it has some (unimportant) values even if u is not a code of a character or if w is not a string. This function can be derived from the function NPB as follows: $y = \mathsf{NOcc}(u, w)$ if and only if there exists a number z not exceeding w such that: the length of the binary expansion of z is $\mathsf{Lh}(w)$, for each x the x-th bit of z is positive if and only if $(w)_x = u$, and $\mathsf{NPB}(z) = y$. Therefore, $y = \mathsf{NOcc}(u, w)$ is a Δ_0-formula. It follows from the summation property of NPB that if w_1 and w_2 are strings, then $\mathsf{NOcc}(u, w_1 * w_2)$ equals $\mathsf{NOcc}(u, w_1) + \mathsf{NOcc}(u, w_2)$.

Using the function NOcc we can define balanced strings, which is the key concept that makes it possible to avoid primitive recursion in logical syntax. It is taken from the part of (Hájek & Pudlák, 1993) written by Pudlák. We again just add some details and implementations to the treatment of syntax in (Hájek & Pudlák, 1993). A string w is *balanced* if its length is at least 2, it contains the same number of left parentheses as right parentheses (in symbols, $\mathsf{NOcc}((, w) = \mathsf{NOcc}(), w))$, and whenever u is a nonempty proper initial segment of w, then $\mathsf{NOcc}((, u) > \mathsf{NOcc}(), u)$. It is clear that a balanced string must start with (and end with). If the concatenation uv of two nonempty strings is balanced, then it follows from the summation property that v contains less left parentheses than right parentheses. *Important property of balanced strings* is this: two balanced substrings u and v of a string w can overlap only if one of them is a substring of the other, and they can start at the same position only if they also end at the same position.

A *quasiterm* is any variable, the (single-element) string 0, or any string of the form $\mathsf{S}(w)$, $+(w)$ or $\cdot(w)$ where (w) is a balanced string. Examples of quasiterms are $+((0))$ and $\mathsf{S}(()()())$. A quasiterm t is a *term* (which we write as $\mathsf{Term}(t)$) if whenever (w) is a balanced substring of t, then either (w) is immediately preceded by the letter S and w is a quasiterm, or it is immediately preceded by $+$ or $\cdot$ and w has the form u, v (notice the typewriter comma) where u and v are quasiterms. $\mathsf{Term}(t)$ is a Δ_0-formula.

Notice that we write the binary symbols in front of their operands, not between them. To have at least one full proof, let us verify that if t and s are terms and u is the string $+(t, s)$, then u is a term. It is clear that a term contains the same number of left parentheses as right parentheses. Therefore, u contains the same number of left parentheses as right parentheses. Using the summation property of the function NOcc and distinguishing the cases whether the leftmost character of t is $+$, $\cdot$, v or 0, one can conclude that (t, s) is a balanced string. Thus u is a quasiterm. To finish the verification that it is a term, let v be a balanced substring of u. Distinguish the

Do We Need Recursion?

three cases according to whether the leftmost parenthesis of v is inside t, or it is inside s, or it is the leftmost character of (t,s). In the last case it follows from the property of balanced strings that $v = (t,s)$ and thus v is as required in the definition of term (it is preceded by a binary symbol and at the same time it consists of two comma-separated quasiterms surrounded by parentheses). If the leftmost parenthesis of v is inside t, then the property of balanced strings implies that v is a substring of t. Since t is a term, v is as required. If the leftmost parenthesis of v is inside s, then v is as required as well because s is a term. This argument is a proof of a part of claim (b) in the following theorem. The remaining considerations in (b) and (c) are similar. Claim (a) is straightforward, and (d) follows from (a)–(c).

Theorem 1 *(a)* PA $\vdash$ Term(0) & $\forall v(\text{Var}(v) \rightarrow \text{Term}(v))$.
(b) PA $\vdash$ If t and s are terms, then $S(t)$, $+(t,s)$ and $\cdot(t,s)$ are terms.
(c) PA $\vdash$ Every term u either (i) is the single-element string 0, or (ii) is a variable, or (iii) has the form $S(t)$ where t is a term, or (iv) has the form $+(t,s)$ or $\cdot(t,s)$ where t and s are terms. The possibilities (i)–(iv) are mutually exclusive and, in (iv), t and s are uniquely determined.
(d) The formulas Var(x) *and* Term(x) *define the set of all variables and the set of all terms respectively.*

We see that exactly the same that was said in the previous section about powers of two, about the exponential function and about the function NPB can be said about variables and terms. The formulas Var(x) and Term(x) are Δ_0 and they not only define what they are supposed to define, but, if they are accepted as definitions of variables and terms inside PA, then PA can prove the expected properties of those notions. For example, every nonstandard model $\mathcal{M}$ of PA contains infinitely (unboundedly) many terms, and every term w in $\mathcal{M}$ that is not a variable or the constant 0 can be uniquely decomposed as $w = +(t,s)$ or $w = \cdot(t,s)$ or $w = S(t)$ where t and s are terms. If w is nonstandard and equals $+(t,s)$ or $\cdot(t,s)$, then at least one of t and s must be nonstandard.

The same that we did with variables and terms can be done with the other syntactic notions. Thus we can introduce formulas Fla(z), OccT(v,t), OccF(v,z), Sent(z), FreeSub(v,z,t), SubF(v,z,t,y), and LogAx(z) that are Δ_0 and express that the string z is a formula, the string t is a term and v is a variable occurring in t, the string z is a formula and v is a variable having free occurrences in z, the string z is a sentence, the string t is a term substitutable for v in z, the string y is the result of substituting a term t for v in a formula z and the string z is a logical axiom respectively. These formulas again have the expected properties. For example, the claim every formula z is either atomic, or it can be uniquely decomposed as $\neg z_1$, $\forall v z_1$, $\exists v z_1$, $(z_1 \rightarrow z_2)$, $(z_1 \& z_2)$ or $(z_1 \vee z_2)$ where z_1 and z_2 are formulas and v is a variable can be proved in PA.

4 Conclusions

The use of primitive recursion can sometimes be replaced by the use of Δ_0 conditions and Δ_0-formulas. In particular, the basic syntactic notions in logic and their properties are Δ_0, which is a more accurate result than that they are primitive recursive. This result makes the arithmetization of syntax more natural. Δ_0-formulas that describe terms, formulas, etc. are in fact close to computer programs that would decide about a string whether it is a term, a formula, etc.

References

Bennet, J. H. (1962). *On Spectra* (Dissertation). Princeton University, Princeton, NJ.

Feferman, S. (1960). Arithmetization of metamathematics in a general setting. *Fundamenta Mathematicae, 49*, 35–92.

Hájek, P., & Pudlák, P. (1993). *Metamathematics of First Order Arithmetic*. Springer.

Odifreddi, P. (1989). *Classical Recursion Theory*. North-Holland.

Pudlák, P. (1983). A definition of exponentiation by a bounded arithmetical formula. *Comm. Math. Univ. Carolinae, 24*(4), 667–671.

Švejdar, V. (2020). *Logic: Incompleteness, Complexity, and Necessity*. London: College Publication. (In preparation)

Vítězslav Švejdar
Charles University, Faculty of Arts
The Czech Republic
E-mail: `vitezslav.svejdar@cuni.cz`

A Classical Bimodal Logic with Varying Essences

ANDREW TEDDER[1]

Abstract: Descartes (in)famously held that there are necessary truths which God, due to His freedom in creating the world, could have made false. This view, now often called the *creation doctrine*, has come under criticism for centuries as being confused or incoherent. According to my preferred reading of Descartes' modal metaphysics, he is an essentialist; he takes modal facts to be reducible to facts about essences. This essentialism, along with some of his theological views, lead him to countenance two different kinds of modal fact. In this paper, I consider a variant of the view I attribute to Descartes as motivating a bimodal logic with essences. I present a set of models (built using the tools of frame semantics for classical modal logics) capturing some of the elements of the view, and prove soundness and completeness with respect to an axiomatic presentation, and go on to consider some potential extensions.

Keywords: modal logic, modal metaphysics, essentialism, Descartes

1 Introduction

Descartes (in)famously held the view, often called the *creation doctrine*, that God freely created the eternal truths. The *eternal truths* are a class of *necessary* truths, including truths of mathematics, the laws of nature, and apparently also logical laws.[2] Descartes was apparently committed to the claim that eternal truths are necessarily true, and yet, in holding that God freely created them (or 'freely made them true'), it would appear that Descartes also commits himself to the claim that God could have made them

---

[1] I'd like to thank the organizers of this event and volume, and furthermore the Logica audience for their interesting comments. For helpful discussions of these issues I'd like to thank Jc Beall, Grace Paterson, David Ripley, Lionel Shapiro, and Stewart Shapiro.

[2] Arguably this class includes all necessary truths, but that is a somewhat subtle interpretive point better considered elsewhere. The perennial problems with the creation doctrine arise even if we consider the eternal truths as a subset of the necessary truths, so it doesn't make any difference to my discussion here.

false. Indeed, he explicitly commits himself to the claim that God could have made the eternal truths false in the following passage, in which Descartes responds to a skeptical Mersenne:

> You ask me by what kind of causality God established the eternal truths. I reply: by the same kind of causality as He created all things, that is to say, as their efficient and total cause. For it is certain that *He is the author of the essence of created things no less than of their existence; and their essence is nothing other than the eternal truths.* You ask also what necessitated God to create these truths; and I reply that He was free to make it not true that all radii of the circle are equal – just as free as He was not to create the world. And it is certain that these truths are no more necessarily attached to His essence than are other created things. (Letter to Mersenne, 27 May, 1630 AT 1:152–53, CSMK3, p. 25, my emphasis)[3]

Furthermore, in the following passage, we find Descartes (apparently) drawing the consequence that since God was not necessitated to make a necessary proposition A true, he could have made $\neg A$ true:[4]

> I turn to the difficulty of conceiving how God would have been acting freely and indifferently if He had made it false that the three angles of a triangle were equal to two right angles, or in general that contradictories could not be true together. It is easy to dispel this difficulty by considering that the power of God cannot have any limits ... [which] consideration shows us that God cannot have been determined to make it true that contradictories cannot be true together, and therefore that He could have done the opposite. (To Mesland, 2 May 1644, AT 4:118–119, CSMK3 p. 235)

The problem is that Descartes seems to commit himself to both the claims (1) that eternal truths are necessary and (2) that eternal truths are possibly

---

[3] Throughout, citations of Descartes' correspondence will be to both the Adam and Tannery (AT) edition and the Cottingham, Stoothoff, Murdoch, and Kenny translation, Cottingham, Stoothoff, Murdoch, and Kenny (1991). This latter will be cited as "CSMK3" for convenience.

[4] I'll read "doing the opposite of making A true" as "making $\neg A$ true", in interpreting the following passage. There is some dispute about what *doing the opposite* here amounts to, in particular see (Ishiguro, 1986), but I won't go into that here.

A Classical Bimodal Logic with Varying Essences

false.[5] The latter seems to follow from the claim that God *could* have made the eternal truths false. There is a clear tension between these claims, since we usually take necessity and possibility to be duals in the sense that for a proposition to be necessarily true is just for it to not be possibly false. This doctrine has been puzzling commentators for centuries, with criticisms coming from Descartes' contemporaries (and near contemporaries) such as Mersenne, Gassendi, Spinoza, and Leibniz, and more recently by Geach (1980) and Van Cleve (1994). There have been a number of readings provided in the past few decades, perhaps most famously by Frankfurt (1977), Curley (1984), Ishiguro (1986), Bennett (1994), and Kaufman (2002).

In my view, Descartes is an essentialist, and so takes modal facts to be reducible to facts about the essences. Furthermore, I take it that this essentialism motivates a bimodal metaphysical picture. This picture incorporates *inner* or i-modalities ($\Box$, $\Diamond$) which express facts about *the essences of created things* and *outer* or o-modalities ($\blacksquare$, $\blacklozenge$) which express facts about *God's essence*.[6] This allows the apparent inconsistency in the view to be avoided by treating the instance of 'necessary' in (1) as 'i-necessary' and the instance of 'possibly' in (2) as 'o-possibly'. The eternal truths are necessary in virtue of God creating essences of created things the way He did, but God's nature did not compel Him to create the world in just that way. In an unpublished manuscript Tedder (n.d.) I present and defend a version of the interpretation of Descartes on offer here, so I do not, in this paper, aim to defend the details of this as a reading of Descartes.[7] My aim in this paper is to use some logical tools to develop a formal account of a variation on the essentialist bimodal metaphysical picture I propose.[8]

---

[5]This way of stating the problem is due to Kaufman (2002).

[6]The reason for the hard and fast distinction between God's essence and the essences of created things is partially due to textual considerations – Descartes distinguishes between these in, for instance, the passage from the 27 May 1630 letter to Mersenne quoted earlier – but also due to the fact that while it is, perhaps, easy to get one's head around God having voluntary control over the essences of His creatures, it is much less clear how He could be said to have voluntary control over His own essence, unless He creates Himself. Perhaps this last claim has some cogent theological defense, but I don't take it to be among Descartes' commitments here, and so take it that the distinction drawn between the two kinds of essences as sufficiently justified.

[7]To get a full understanding of Descartes' modal metaphysics, one must reckon with various Cartesian theses about God, such as Descartes' (somewhat idiosyncratic) versions of divine simplicity, omnipotence, etc, and account for a variety of difficulties which arise in understanding the creation doctrine in conjunction with Descartes' various metaphysical views. A proper study of the text is much more involved than can be undertaken here, hence it is left for Tedder (n.d.).

[8]The resulting view isn't really *Cartesian*, but is an, I think, interesting *neo-Cartesian* view.

Andrew Tedder

I'll develop a model theory aimed at both (a) reading modal facts in terms of essences and (b) allowing for worlds at which objects can have different essences than they actually do. The rough picture suggested by (a) and (b) is of a space of possible worlds split into equivalence classes (cells), where each object in a world has all the same essential properties as it (or its counterpart, if you prefer) does in all worlds in the same cell. Call any two worlds standing in this relation *essentially similar*. We can state truth conditions, in terms of a satisfaction relation ⊩ (to be fleshed out momentarily) for a pair of pairs of modal operators as follows:

- $\alpha \Vdash \Box A$ iff for all β which are essentially similar to α, $\beta \Vdash A$.
- $\alpha \Vdash \Diamond A$ iff for some β which is essentially similar to α, $\beta \Vdash A$.
- $\alpha \Vdash \blacksquare A$ iff for all β, $\beta \Vdash A$.
- $\alpha \Vdash \blacklozenge A$ iff for some β, $\beta \Vdash A$.

The bimodalism is represented by having local and global modalities, and I'll develop some formal machinery allowing me to distinguish essential from non-essential predications of objects. To pull it all together, the aim is to find constraints on models appropriate to ensure that the accessibility relation interpreting the i-modalities does indeed coincide with essential similarity. This amounts to finding a set of models expressing the essentialism and its relation to the bimodalism.[9] I'll proceed using the toolkit of classical modal logic and provide a frame-based semantics with some of these features and with a simple axiomatisation. I'll go on to narrow the class of models to better account for the essentialism. I'll end by considering how this class of models can be altered to account for more of the interesting consequences of the creation doctrine, when put into this framework.

1.1 The Language

We begin with a zero-order language (incorporating some structure for atomic formulas, but no quantifiers) built from a signature including:

- $\mathcal{N}$ a set of name constant letters $a_1, \ldots, a_n, \ldots$.

Perhaps the most substantial departure from Descartes, and the most important for my purposes, is that I make use of possible worlds. I don't think Descartes is committed to the existence of possible worlds (or even that his view is best understood in such terms), but that something like his metaphysical view can be helpfully understood in those terms.

[9]The o-modalities can be understood just as a global modalities, seeing all worlds and, hence, all essences which might be assigned to any object at any world.

A Classical Bimodal Logic with Varying Essences

- $\mathcal{P}$ a set of predicate letters, and $ar : \mathcal{P} \longrightarrow \mathbb{N}$, a function assigning arities to predicate letters.

In constructing the language, $\cdot^\Box$ is an operation taking unary predicate letters to *fresh* unary predicate letters, i.e., predicate letters not occurring in $\mathcal{P}$. Fix an enumeration of the unary predicates $P_1, \ldots, P_i, \ldots$ and read $P_1^\Box, \ldots, P_i^\Box, \ldots$ as the essentialised versions of these predicates: i.e., $P_i^\Box a$ should be understood to mean "a is essentially P_i". Let $\mathcal{P}^\Box$ be the set of these $P_i^\Box$, and let $\mathcal{P}^* = \mathcal{P} \cup \mathcal{P}^\Box$.

Definition 1 *$\mathcal{L}$ is defined recursively as follows:*

- $t \in \mathcal{L}$, t is a truth constant[10]
- *If $R \in \mathcal{P}^*$, $ar(R) = n$, and $a_1, \ldots, a_n \in \mathcal{N}$ then $R(a_1, \ldots, a_n) \in \mathcal{L}$.*
- *$A \in \mathcal{L}$ only if $\neg A, \Box A, \blacksquare A \in \mathcal{L}$.*
- *$A, B \in \mathcal{L}$ only if $A \wedge B, A \vee B \in \mathcal{L}$.*

In addition to these are the defined connectives $A \supset B := \neg A \vee B$ and $A \equiv B := (A \supset B) \wedge (B \supset A)$.

2 VE Frames

All models considered here are built on a common core of frames:

Definition 2 *A **VE** frame $\mathcal{F}$ is a tuple $\langle W, N, R, S, D, e, \{X_i \mid i \in \mathbb{N}\}\rangle$ where:*

- $W \neq \varnothing$ *is a set of worlds $\{\alpha, \beta, \gamma, \ldots\}$*
- $N \subseteq W$ *and* $N \neq \varnothing$
- $R, S \subseteq W^2$
- $D \neq \varnothing$ *is a denumerable set of objects $\{\bar{a}_1, \ldots, \bar{a}_n \ldots\}$*
- $e : W \times D \longrightarrow \wp\wp(D)$
- $X_i : W \longrightarrow \wp(D)$

---

[10] t is added for technical convenience, but also to aid us in thinking of N as the set of actually essentially possible worlds—those worlds at which all actually i-necessary truths are true. This is just part of a richer language one might want to add in order to formalise more complex propositions, and which I intend to add in future work, but for now I'll stick with t by itself.

This includes a space of worlds W, and a distinguished set of normal worlds N, two binary accessibility relations on W, a domain D, and finally a set of properties in intension $\{X_i\}$, and a function e assigning to each object at each world a set of properties (intuitively, those properties essential to that object at that world). On $\mathcal{F}$ we place the following constraints (where $\alpha \in W$, let $R\alpha = \{\beta \mid R\alpha\beta\}$):[11]

(c1) R, S are equivalence relations

(c2) $S = W^2$

(c3) $X_i\alpha \in e(\alpha, \bar{a}) \Rightarrow \forall \beta \in R\alpha(X_i\beta \in e(\beta, \bar{a}))$

(c4) $X_i\alpha \in e(\alpha, \bar{a}) \Rightarrow \forall \beta \in R\alpha(\bar{a} \in X_i\beta)$

(c5) $\alpha \in N \,\&\, R\alpha\beta. \Rightarrow \beta \in N$

(c6) $\alpha, \beta \in N \Rightarrow R\alpha\beta$

In considering these constraints, let's take a moment to further fill in the picture of modal space the *bimodal, essentialist* view attributes to Descartes. First, with regard to the *bimodal* part, the space of worlds ($S = W^2$) is understood as coming in *cells* – equivalence classes w.r.t. R – and the set of normal worlds N is an R-cell. Intuitively, N can be understood as the worlds R-accessible to the actual world (though the actual world itself doesn't feature explicitly in the models).

Second, given the intuitive reading of e and X_i, when, in a model, $X_i\alpha \in e(\alpha, \bar{a})$ holds, this represents the fact that the property (in extension) picked out by $X_i\alpha$ is essential to $\bar{a}$ at α. With this in mind, (c3) expresses the fact that if a property is essential to $\bar{a}$ at α then it is essential to $\bar{a}$ at any R-accessible β. In short: essential predications are i-necessary. (c4) expresses the claim that when P_i is an essential property of something, then is is also an i-necessary property of that thing. (c3) relates two worlds being R-related to their being essentially similar, as it has the consequence that if $R\alpha\beta$ then for any object $\bar{a}$, $e(\alpha, \bar{a}) = e(\beta, \bar{a})$.

3 Classical Variable Essences: CVE

Definition 3 (**CVE** model) M is a **CVE** model when $M = \langle \mathcal{F}; g, f \rangle$ where:

---

[11] Instead of enforcing (c2), I could instead have just required that $S = W^2$ in the definition of **VE**-frame. I have done it this way for the sake of explicitness in the canonical model argument to come—so that (c2) is just one more condition to check in the appropriate lemma.

A Classical Bimodal Logic with Varying Essences

- $\mathcal{F}$ is a **VE** frame
- $g : \mathcal{N} \longrightarrow D$
- When Q is an n-place predicate, $f(Q)$ is a function of type $W \longrightarrow \wp(D^n)$. For unary predicates P_i, set $f(P_i) = X_i$.

Note that including the constraint that $f(P_i) = X_i$ means that when I obtain completeness results below, they hold with respect to a restricted set of models on **VE** frames, and not to *all* such models. This is a byproduct of the syntactically flavoured approach to representing essential properties I adopt here, but it does limit the capacity for this set of models to be studied algebraically, and for many classical methods for studying modal logics to be directly applied to the resulting system. I leave it to the reader to judge whether the tradeoff is worth it.[12]

Satisfaction, $\Vdash \,\subseteq W \times \mathcal{L}$, is defined as follows, where A, B are metavariables over $\mathcal{L}$:

- $\alpha \Vdash t \iff \alpha \in N$
- $\alpha \Vdash Pa_1,\ldots,a_n \iff \langle g(a_1),\ldots,g(a_n) \rangle \in f(P)\alpha$
- $\alpha \Vdash P_i^\square a \iff f(P_i)\alpha \in e(\alpha, g(a))$
- $\alpha \Vdash \neg A \iff \alpha \nVdash A$
- $\alpha \Vdash A \wedge B \iff \alpha \Vdash A \,\&\, \beta \Vdash B$
- $\alpha \Vdash A \vee B \iff \alpha \Vdash A \text{ or } \alpha \Vdash B$
- $\alpha \Vdash \square A \iff \forall \beta \in R\alpha(\beta \Vdash A)$
- $\alpha \Vdash \Diamond A \iff \exists \beta \in R\alpha(\beta \Vdash A)$
- $\alpha \Vdash \blacksquare A \iff \forall \beta \in S\alpha(\beta \Vdash A)$
- $\alpha \Vdash \blacklozenge A \iff \exists \beta \in S\alpha(\beta \Vdash A)$

Definition 4 *Given a model* M, $\Gamma \vDash_M A$ *iff* $\forall \alpha \in N^\mathcal{M}(\forall G \in \Gamma(\alpha \Vdash G) \Rightarrow \alpha \Vdash A)$.

Definition 5 $\Gamma \vDash_{CVE} A$ *iff for every* **CVE** *model* M, $\Gamma \vDash_M A$.

---
[12]This might lead some to claim that the system studied here is not, in fact, a *logic*. I encourage those of that inclination to consider this system a model of an interesting position in modal metaphysics (and its commitments) rather than a theory of logical consequence as it concerns modal operators.

3.1 CVE: A Hilbert System

As before, A, B are metavariables over $\mathcal{L}$, and furthermore the following are defined connectives: $A \supset B := \neg A \vee B$ and $A \equiv B := (A \supset B) \wedge (B \supset A)$. Below, in (A1)–(A5), $\langle \otimes, \oplus \rangle$ is either $\langle \Box, \Diamond \rangle$ or $\langle \blacksquare, \blacklozenge \rangle$.

Axioms:

A0 All tautologies of zero-order Classical logic

A1 $\oplus A \equiv \neg \otimes \neg A$

A2 $\otimes(A \supset B) \supset (\otimes A \supset \otimes B)$

A3 $\otimes A \supset A$

A4 $\otimes A \supset \otimes \otimes A$

A5 $\oplus A \supset \otimes \oplus A$

A6 $\Diamond A \supset \blacklozenge A$

A7* $P_i^\Box a \supset \Box P_i^\Box a$

A8* $P_i^\Box a \supset \Box P_i a$

A9 t

A10 $t \supset \Box t$

A11 $(t \wedge \Box A) \supset \blacksquare(t \supset A)$

Rules:

R1 $A \supset B, A \Rightarrow B$

R2 $A \Rightarrow \Box A$

Note these are rules of proof, so '$A_1, \ldots, A_n \Rightarrow B$' is to be read 'when $A_1, \ldots, A_n$ are theorems, then so is B.'

The $*$ marks a restriction on A7, A8 – they may only be instantiated by by essentialised-predicates in the antecedents, and not just by any predicate. So one cannot, for instance, substitute $P \notin \mathcal{P}^\Box$ for $P_i^\Box$ in A8 to obtain $Pa \supset \Box Pa$.

Definition 6 (**CVE** Proof) *A **CVE** proof is a finite series of formulae, each of which is either an assumption, an instance of a **CVE** axiom, or results from the application of one of the **CVE** rules to previous formulae occurring in the series.*

Definition 7 (**CVE** Theorem/Consequence) *$\vdash_{CVE} A$ iff there is a **CVE** proof of which A is the last formula. $\Gamma \vdash_{CVE} A$ iff there is a **CVE** proof of A from formulas in Γ taken as premises.*

3.2 CVE Adequacy

Theorem 1 (Soundness) $\Gamma \vdash_{CVE} A \Rightarrow \Gamma \vDash_{CVE} A$

A Classical Bimodal Logic with Varying Essences

Proof. The proof is, as usual, by induction. I'll omit the standard cases, and present just a few cases involving the added vocabulary.

(A7) Suppose that $\alpha \in N$ and that $\alpha \Vdash P_i^\square a$. That is, $f(P_i)\alpha \in e(\alpha, g(a))$, and so $X_i\alpha \in e(\alpha, g(a))$. By (c3), $\forall \beta \in R\alpha(X_i\beta \in e(\beta, g(a)))$, and thus $f(P_i)\beta \in e(\beta, g(a))$. By the truth condition for $P_i^\square$, $\forall \beta \in R\alpha(\beta \Vdash P_i^\square a)$, and by the condition for $\square$, $\alpha \Vdash \square P_i^\square a$.

(A8) Suppose that $\alpha \in N$ and that $\alpha \Vdash P_i^\square a$. It follows that $f(P_i)\alpha = X_i\alpha \in e(\alpha, g(a))$ and so, by (c4), $\forall \beta \in R\alpha(g(a) \in X_i\beta = f(P_i)\beta)$. Thus $\forall \beta \in R\alpha(\beta \Vdash P_i a)$, and so $\alpha \Vdash \square P_i^1 a$.

(A11) Suppose that $\alpha \in N$, $\alpha \Vdash t \wedge \square A$, and furthermore $\beta \Vdash t$. Then $\beta \in N$, and so by (c6), $R\alpha\beta$, so $\beta \Vdash A$, and thus $\beta \Vdash t \supset A$. Since β was arbitrary, $\alpha \Vdash \blacksquare(t \supset A)$, as desired. $\square$

For the completeness proof, the method is the standard canonical model method with one twist to deal with the global modality ■ (see Blackburn, de Rijke, & Venema, 2001, pp. 417–418 for details). I start by defining a pre-canonical model, which satisfies all conditions on **CVE** models except (C2) and (C6), and then we obtain the actual canonical model from this pre-canonical model which, in addition, satisfies these constraints.

Definition 8 (**CVE** Pre-Canonical Model) *The **CVE** pre-canonical model PM_c is a tuple $\langle W_c, N_c, R_c, S_c, D_c, e_c, \{X_{i_c}\}, f_c, g_c \rangle$ where:*

- $W_c = \{\alpha \subseteq \mathcal{L} \mid \alpha \text{ is maximally consistent and } \vdash_{\mathbf{CVE}} A \supset B, A \in \alpha. \Rightarrow B \in \alpha\}$
- $N_c = \{\alpha \in W_c \mid t \in \alpha\}$
- $R_c = \{\langle \alpha, \beta \rangle \in W_c^2 \mid \square A \in x \Rightarrow A \in y\}$
- $S_c = \{\langle \alpha, \beta \rangle \in W_c^2 \mid \blacksquare A \in x \Rightarrow A \in y\}$
- $D_c = \mathcal{N}$
- $X_{i_c}\alpha = \{a \in D_c \mid P_i a \in \alpha\}$
- $e_c(\alpha, a) = \{X_{i_c}\alpha \mid P_i^\square a \in \alpha\}$
- $f_c(P_i^n)\alpha = \{\langle a_1, \ldots, a_n \rangle \in D_c^n \mid P_i^n a_1, \ldots, a_n \in \alpha\}$
- $g_c(a) = a$

Lemma 1 (Truth in PM_c) $\quad \alpha \Vdash_c A \iff A \in \alpha$

Proof. The proof proceeds by induction. The cases for $\wedge, \neg, \Box$ are standard, and that for $\blacksquare$ closely mimics that of $\Box$, so I'll only sketch the proof for $\blacksquare$. The remaining work mostly concerns the base case, which is straightforward. As a preliminary to this, note that it follows immediately from the definition that for every unary predicate P_i, $f_c(P_i)\alpha = X_{i_c}\alpha$ for every $\alpha \in W_c$.

Base: Let A be atomic. That is, either A is of the form $P_i^n a_1, \ldots, a_n$, $P_i^\Box a$, or is t. The former is immediate, and note that $\alpha \Vdash_c P_i^\Box a$ iff $f_c(P_i^1)\alpha = X_{i_c}\alpha \in e_c(\alpha, a)$ iff $P_i^\Box a \in \alpha$. For t, the result holds by definition.

Case: ($\blacksquare$) Suppose that $\blacksquare A \in \alpha$ and $S_c \alpha \beta$. It follows that $A \in \beta$, and so by IH $\beta \Vdash_c A$, thus $\alpha \Vdash_c \blacksquare A$. Suppose that $\blacksquare A \notin \alpha$ – it is easy to show that in this circumstance $\exists \beta (S_c \alpha \beta \,\&\, A \notin \beta)$, and thus with another use of IH, it follows that $\alpha \nVdash_c \blacksquare A$. $\square$

Lemma 2 *PM_c satisfies (c1) and (c3)–(c5) of the definition of **VE** frame, and, in addition, $R_c \subseteq S_c$.*

Proof. The proof proceeds by cases. (c1) is straightforward (following the usual procedure) and so is omitted.

(c3) Suppose that $\alpha \in W_c$, $a \in D_c$, $X_{i_c}\alpha \in e_c(\alpha, a)$, and $R_c \alpha \beta$. Since $X_{i_c}\alpha \in e_c(\alpha, a)$, $P_i^\Box a \in \alpha$. $\vdash_{\mathbf{CVE}} P_i^\Box a \supset \Box P_i^\Box a$, so $\Box P_i^\Box a \in \alpha$, and since $R_c \alpha \beta$, $P_i^\Box a \in \beta$. Thus $X_{i_c}\beta \in e_c(\beta, a)$ as desired – this follows from the definition of e_c, X_{i_c}.

(c4) Suppose that $\alpha \in W_c, a \in D_c, X_{i_c}\alpha \in e_c(\alpha, a)$, and $R_c \alpha \beta$. As before, we have $P_i^\Box a \in \alpha$, and since $\vdash_{\mathbf{CVE}} P_i^\Box a \supset \Box P_i a$, it follows that $\Box P_i a \in \alpha$, and since $R_c \alpha \beta$, $P_i a \in \beta$. Thus, $a \in f_c(P_i)\beta$, and so $a \in X_{i_c}\beta$, since $f_c(P_i)\beta = X_{i_c}\beta$.

(c5) Suppose $R_c \alpha \beta$, $\alpha \in N$. Then $t \in \alpha$, and so $\Box t \in \alpha$ as $\vdash_{\mathbf{CVE}} t \supset \Box t$, so $t \in \beta$, so $\beta \in N$.

That $R_c \subseteq S_c$ is guaranteed by the fact that $\vdash_{\mathbf{CVE}} \Diamond A \supset \blacklozenge A$. $\square$

Definition 9 *Let $\alpha \in W_c$. $M_c^\alpha = \langle W_c^\alpha, N_c^\alpha, R_c^\alpha, S_c, D_c^\alpha, e_c^\alpha, \{X_{i_c}^\alpha\}, f_c^\alpha, g_c^\alpha \rangle$ where:*

- $W_c^\alpha = S_c \alpha$
- $N_c^\alpha = \{\beta \in W_c^\alpha \mid t \in \beta\}$

- $R_c^\alpha = R_c \upharpoonright W_c^\alpha$
- $D_c^\alpha = D_c$
- $e_c^\alpha(-, a) = \{e_c(\beta, a) \mid \beta \in W_c^\alpha\}$

The $X_{i_c}^\alpha$s, f_c^α, g_c^α, and $\Vdash_c^\alpha$ are defined similarly to e_c^α as the restrictions of the originals to W_c^α.

Lemma 3 *For all $\alpha \in W_c$, M_c^α is a **CVE** Model.*

Proof. First note that the desired truth conditions hold in M_c^α for $\Vdash_c$ follows directly from the definitions of PM_c, M_c^α, and Lemma 1. In M_c^α we have that $S_c = W_c^\alpha$, by the construction, so M_c^α satisfies (c2). Furthermore, M_c^α satisfies (c6), for we have that $\blacksquare A \in \gamma \iff \forall \beta \in W_c^\alpha (A \in \gamma)$ for any $\gamma \in W_c^\alpha$, so if $t \in \alpha, \beta$ and $\Box A \in \alpha$, then $\blacksquare(t \supset A) \in \alpha$ since $\vdash (t \wedge \Box A) \supset \blacksquare(t \supset A)$, and thus $t \supset A \in \beta$ and so $A \in \beta$, and thus $R_c \alpha \beta$. All other conditions were satisfied by PM_c, and all are bounded universal generalisations, and so remain true in M_c^α.

It remains to show that $N_c^\alpha \neq \emptyset$. This is easily done by a standard Lindenbaum construction. Let $\alpha_0' = \{t\} \cup \{A \mid \blacksquare A \in \alpha\}$. Note that α_0' is consistent because $\vdash_{\mathbf{CVE}} \blacksquare \neg t \supset A$ holds for every A and $\alpha \in W_c$. Next, enumerate $\mathcal{L}$ as $B_1, \ldots, B_n, \ldots$, and let:

$$\alpha_{n+1}' = \begin{cases} \alpha_n' \cup \{B_n\} & \text{if the result is consistent} \\ \alpha_n' \cup \{\neg B_n\} & \text{else} \end{cases} \quad (1)$$

and $\alpha' = \bigcup_{n \in \omega} \alpha_n'$. This guarantees that $\alpha' \in N_c$ (furthermore $S_c \alpha \alpha'$, so $\alpha' \in N_c^\alpha$). □

Theorem 2 $\Gamma \vDash_{\mathbf{CVE}} A \Rightarrow \Gamma \vdash_{\mathbf{CVE}} A$

Proof. Suppose that $\Gamma \nvdash_{\mathbf{CVE}} A$. So $\Gamma \cup \{\neg A\}$ is consistent, and furthermore $\neg t \notin \Gamma$ (note that $\neg t \nvdash_{\mathbf{CVE}} A$). Then let:

$$\alpha_0 = \{B \mid \exists A_1, \ldots, A_n \in \Gamma \cup \{\neg A, t\} (\vdash_{\mathbf{CVE}} \bigwedge_{1 \leq i \leq n} A_i \supset B)\}$$

Note that if $\vdash_{\mathbf{CVE}} A$ then $A \in \alpha_0$, for if $\vdash A$ then $\vdash t \supset A$. Since $\Gamma \cup \{\neg A, t\}$ is consistent, so is α_0. We can extend it to an element of $\alpha \in N_c$ as in Lemma 3. As before this guarantees that $\alpha \in N_c$. We can generate M_c^α as above, and note that in this model $\Gamma \subseteq \alpha$ and yet $A \notin \alpha$, so $\Gamma \nvDash_{M_c^\alpha} A$, and thus $\Gamma \nvDash_{\mathbf{CVE}} A$. □

3.3 The Key CVE Consequence

Theorem 3 $P_i^\Box a \wedge \blacklozenge \neg P_i a$ *is satisfiable. Hence* $P_i^\Box a \supset \blacksquare P_i a$ *is not a theorem of* **CVE**.

Proof. For simplicitly, let the language signature consist of $\mathcal{P} = \{P_i\}$ with $ar(P_i) = 1$ and $\mathcal{N} = \{a\}$. Let M be a **CVE** model with the following elements:

$W = \{@, \alpha\}$ $\qquad\qquad f(P_i^1)@ = X_i@ = \{\overline{a}\}$
$N = \{@\}$ $\qquad\qquad\qquad f(P_i^1)\alpha = X_i\alpha = \varnothing$
$R = \{\langle @, @\rangle, \langle \alpha, \alpha\rangle\}$ $\qquad g(a) = \overline{a}$
$S = W^2$ $\qquad\qquad\qquad e(@, \overline{a}) = \{X_i@\}$
$D = \{\overline{a}\}$ $\qquad\qquad\qquad e(\alpha, \overline{a}) = \varnothing$

On this model, $@ \Vdash P_i^\Box a$, $\alpha \Vdash \neg P_i a$. Hence $\vDash_M P_i^\Box a \wedge \blacklozenge \neg P_i a$. $\qquad\square$

Corollary 1 $\Box A \supset \blacksquare A$ *is not a theorem of* **CVE**.

This corollary indicates that we can build models where $\blacksquare$ behaves as it should for some formulae, but the **C** in **CVE** nevertheless does hold us back from the account we really want, as for any classical tautology A, we have $\vdash_{\mathbf{CVE}} \blacksquare A$. This does not settle well with the target interpretation, according to which even logical truths are o-possibly false.

Even putting aside the need for non-classicality to complete this picture, we are also missing another desirable component: we want that if α, β are essentially similar, they are i-accessible to one another. This way, we really will be *analysing* the i-accessibility of worlds, and hence their i-modal properties, in terms of essential similarity. To that end, we need to beef up the expressive machinery available.

4 Extending CVE

The key element of the *essentialist* part of the interpretation can be expressed formally as an analysis as follows:

(∗) $\qquad \forall \overline{a} \in D \forall X_i (X_i \alpha \in e(\alpha, \overline{a}) \iff X_i \beta \in e(\beta, \overline{a})) \iff R\alpha\beta$

expressing the claim that co-i-possibility is essential similarity. The right to left direction follows immediately from (c3), but for the converse direction more is needed to obtain a proof system. A way to proceed is by adding

A Classical Bimodal Logic with Varying Essences

further expressive resources – namely, we could add to the language an operator binding a formula to a particular world (a hybrid operator), where for all $\alpha, \beta \in W$:

$$\alpha \Vdash @_\beta A \iff \beta \Vdash A$$

With this, the natural candidate defining the left to right half of the above is the infinitary rule:

(Ess) $\{@_\alpha P_i^\Box a \equiv @_\beta P_i^\Box a \mid P_i^\Box a \in \mathcal{L}\} \Rightarrow @_\alpha \Box A \supset @_\beta A$

To make this work even in a preliminary way, we need one further assumption in the definition of model, namely, that g is a surjection, (every $\bar{a} \in D$ is the referent of some $a \in \mathcal{N}$).

Proposition 1 *If* $(*)$ *is satisfied in* M, *and* g^M *is a surjection, then every world* $\alpha \in W^M$ *is closed under the rule (Ess).*

Proof. Fix $\gamma \in W^M$, and suppose that for every $P_i^\Box, a, \alpha, \beta$,

$$\gamma \Vdash @_\alpha P_i^\Box a \equiv @_\beta P_i^\Box a$$

It follows by the truth conditions for $\equiv$, and $@_\alpha, @_\beta$ that:

$$\alpha \Vdash P_i^\Box a \iff \beta \Vdash P_i^\Box a$$

So, since $f(P_i) = X_i$, and since g^M is surjective, it follows that for every $X_i, \bar{a} \in D$:

$$X_i \alpha \in e(\alpha, \bar{a}) \iff X_i \beta \in e(\beta, \bar{a})$$

and thus $R\alpha\beta$, by $(*)$. Since $\gamma \Vdash @_\alpha \Box A$, $\alpha \Vdash \Box A$ and so $\beta \Vdash A$, and thus $\gamma \Vdash @_\beta A$. Thus $\gamma \Vdash @_\alpha \Box A \supset @_\beta A$. □

So (Ess) is sound in the presence of $(*)$, but completeness is another matter, and I'm not in a position to give a completeness proof here. In lieu of that, consider the following suggestive proposition, establishing that under some pretty general assumptions concerning the language signature, that (Ess) is counterexemplified in some **CVE** models on frames failing to satisfy $(*)$.

Proposition 2 *In any language with at least one unary and at least one binary predicate, if a frame* F *does not satisfy* $(*)$ *then there is a model (satisfying the constraints on* **CVE** *model-hood) on* F *which invalidates (Ess).*

Andrew Tedder

Proof. Set the language as follows (for simplicity): $\mathcal{N} = \{a\}, \mathcal{P} = \{P, R\}$ with $ar(P) = 1, ar(R) = 2$. So an appropriate F for the language will have for every $\alpha \in W^F$, $X\alpha = f(P)\alpha$. Furthermore, in supposing that F does not satisfy $(*)$, we must have that for some $\alpha, \beta \in W^F, \bar{a} \in D^F$, we have that $X\alpha \in e(\alpha, \bar{a}) \iff X\beta \in e(\alpha, \bar{a})$ but $\langle \alpha, \beta \rangle \notin R$. Since g is assumed to be a surjection, we have for every $\bar{a}$, there is some a s.t. $g(a) = \bar{a}$. Let M also be such that $f(R)\alpha = D^2$, $f(R)\beta = \varnothing$, and $f(R)\gamma = D^2$ for any γ s.t. $R\alpha\gamma$, and $f(R)\delta = \varnothing$ for every other world. So, by these suppositions we have:

- $\alpha \Vdash P^\square a \iff \beta \Vdash P^\square a$
- $\alpha \Vdash \square R(a, a)$
- $\beta \nVdash R(a, a)$

Since R is an equivalence, we be can be sure that there is no δ s.t. $R\alpha\delta$ and $R\delta\beta$. So it is safe to define $f(R)$ as above. So $R(a, a)$ is satisfied by all and only worlds accessible from α along R. Furthermore, by the truth conditions for the hybrid operators, for any $\gamma \in N^M$:

- $\gamma \Vdash @_\alpha P^\square a \equiv @_\beta P^\square a$
- $\gamma \Vdash @_\alpha \square R(a, a)$
- $\gamma \nVdash @_\beta R(a, a)$

Thus $\gamma \nVdash @_\alpha \square R(a, a) \supset @_\beta R(a, a)$, but for all $P^\square, a$, $\gamma \Vdash @_\alpha P^\square a \equiv @_\beta P^\square a$, as desired. $\square$

In the case where $f(P)\alpha \notin e(\alpha, \bar{a})$ and $f(P)\beta \notin e(\alpha, \bar{a})$, we can even set that $\bar{a} \in f(P)\delta$ for all $\delta \in R\alpha$, and $\bar{a} \notin f(P)\beta$, to obtain $\alpha \Vdash \square Pa$ and $\beta \nVdash Pa$ for the counterexample.

These considerations suggest an infinitary proof system appropriate for the target classical models. This would involve adding axioms for the relativised actuality operators, as well as whatever other hybrid vocabulary is needed to completely axiomatise the resulting modal logic (for instance, by adapting the axioms of Blackburn et al. (2001)[§7.3]), and finally adding to this (Ess) and adapting the canonical model argument given above to incorporate *labeled* maximally consistent sets.

A Classical Bimodal Logic with Varying Essences

5 Conclusion and Future Directions

I have defined two sets of models, one a subset of the other, which capture key elements of a bimodal, essentialist view inspired by Descartes' creation doctrine. These models are (relatively) simple, and flesh out part of the structure of the target metaphysical position, as I understand it.

However they only capture part of the target view, and the most nicely behaved part at that, so the models developed here only tell part of the story. In order to fill in the picture, some additional machinery must be added to allow for worlds God could have created in which contradictories can be true together, for instance. On my understanding of the creation doctrine, one should take commitment to the claim "God could have made contradictories true together" as implying "God could have made the logical vocabulary behave in any way." On a contemporary understanding of logic, this last claim can be cashed out as 'God could have made the logical consequence behave in any way.' So a fuller version of the model structure I present here would include variation among worlds of the behaviour of the logical vocabulary, as well as variation among essences.[13] This suggests an approach following on work by Rantala (1982) on what Priest (2006) calls "open worlds."

In addition, it would be interesting to adapt the more complex treatment of essences in the work of Fine (1995) to the present view. While the treatment of essences I use here is simple, the fact that we must specify features of models and not just of frames renders the view less general than would be desirable, and taking on the additional complexity of Fine's treatment may provide room to solve that problem. Indeed, in order to fully capture the reductionism in the essentialism discussed here, some substantial complexities are needed, as discussed in Section 4, so the apparent simplicity may only be available in a relatively small fragment of the target system.

---

[13] It is important to note that this discussion of logic involves understanding logic and inference in a way quite alien to Descartes himself—see (Gaukroger, 1989) for a nice discussion of how Descartes actually understood inference. While I hold that on Descartes' own view, the claim that God could have made logic behave in any way does follow from the creation doctrine, capturing what this would mean *for Descartes* would require doing something other than what I am suggesting here. My proposal relies on a contemporary understanding of logic, and this is another sense in which what is being modeled here is not Descartes, but an interesting neo-Cartesian view. It seems to me that such a view is potentially interesting as a variation on recent views arguing that logic is, in some sense, contingent (for instance by Mortensen, 1989 and Sandgren & Tanaka, in press), even if it is not directly applicable as an account of Descartes' own commitments.

References

Bennett, J. (1994). Descartes's theory of modality. *The Philosophical Review*, *103*(4), 639–667.
Blackburn, P., de Rijke, M., & Venema, Y. (2001). *Modal Logic*. Cambridge University Press.
Cottingham, J., Stoothoff, R., Murdoch, D., & Kenny, A. (1991). *The Philosophical Writings of Descartes* (Vol. 3). Cambridge University Press.
Curley, E. M. (1984). Descartes on the creation of the eternal truths. *The Philosophical Review*, *93*(4), 569–597.
Fine, K. (1995). The Logic of Essence. *Journal of Philosophical Logic*, *24*(3), 241–273. (Special Issue: Afterthoughts on Kaplan's Demonstratives)
Frankfurt, H. (1977). Descartes on the creation of the eternal truths. *The Philosophical Review*, *86*(1), 36–57.
Gaukroger, S. (1989). *Cartesian Logic: An Essay on Descartes's Conception of Inference*. Oxford University Press.
Geach, P. (1980). *Logic Matters*. University of California Press.
Ishiguro, H. (1986). The status of necessity and impossibility in Descartes. In A. O. Rorty (Ed.), *Essays on Descartes' Meditations* (pp. 459–471). University of California Press.
Kaufman, D. (2002). Descartes' creation doctrine and modality. *Australasian Journal of Philosophy*, *80*(1), 24–41.
Mortensen, C. (1989). Anything is possible. *Erkenntnis*, *30*(3), 319–337.
Priest, G. (2006). *In Contradiction: A Study of the Transconsistent* (2nd ed.). Oxford University Press.
Rantala, V. (1982). Quantified modal logic: Non-normal worlds and propositional attitudes. *Studia Logica*, *41*(1), 41–65.
Sandgren, A., & Tanaka, K. (in press). Two kinds of logical impossibility. *Noûs*. doi: 10.1111/nous.12281
Tedder, A. (n.d.). *An Essentialist Bimodal Interpretation of Descartes' Creation Doctrine*. (Manuscript)
Van Cleve, J. (1994). Descartes and the destruction of the eternal truths. *Ratio*, *7*, 58–62.

Andrew Tedder
Czech Academy of Sciences, Institute of Computer Science
The Czech Republic
E-mail: `ajtedder.at@gmail.com`

Categoricity and Possibility. A Note on Williamson's Modal Monism

IULIAN D. TOADER

Abstract: The paper sketches an argument against modal monism, more specifically against the reduction of physical possibility to metaphysical possibility. The argument is based on the non-categoricity of quantum logic.

Keywords: modal monism, quantum logic, non-categoricity

1 Modal monism and anti-exceptionalism

A widespread philosophical view about modality holds that there exists only one kind of necessity and possibility, to which all other kinds are to be reduced. The view is often called *modal monism*, and it has been expressed perhaps most famously by Wittgenstein in the *Tractatus*: "The only necessity that exists is logical necessity" (6.37). "Just as the only necessity that exists is logical necessity, so too the only impossibility that exists is logical impossibility" (6.375). According to this view, that is physically possible which is logically consistent with the laws of nature.

A similar view, according to which metaphysical and physical modality "stand or fall" together, has been more recently defended by Timothy Williamson (2016): "In given circumstances, a proposition is *nomically possible* if and only if it is metaphysically compossible with what, in those circumstances, are the laws of nature (their conjunction is metaphysically possible)" (p. 455). Thus, nomical or physical possibility cannot be explained independently of metaphysical possibility.

It seems natural to take modal monism to provide support to Williamson's *anti-exceptionalism* about metaphysics, in general, and about the metaphysics of modality, in particular, which is expressed in the following way: "We should not treat the metaphysics and epistemology of metaphysical modality in isolation from the metaphysics and epistemology of the natural sciences." (Williamson, 2016, p. 453). For if physical modality is explained in terms of metaphysical modality, then this anti-exceptionalist requirement is arguably satisfied.

Iulian D. Toader

In an earlier book, wherein Williamson mounts a defense of a metaphysical view he calls *necessitism*, i.e., the view that necessarily everything is such that necessarily something is identical with it, he similarly affirms anti-exceptionalism, which is here expressed somewhat differently: "We will be guided throughout this book by a conception of theories in logic and metaphysics as scientific theories, to be assessed by the same overall standards as theories in other branches of science" (Williamson, 2013, p. 27).[1] These anti-exceptionalist declarations are meant to stand on their own, with no support from modal monism, which does not seem to be part of the picture yet: "Metaphysical possibility will not be assumed to be metaphysically basic, or fundamental, or irreducible, or perfectly natural, or anything like that" (Williamson, 2013, p. 3, footnote 4).

Adopting scientific standards for the assessment of the metaphysics of metaphysical modality might be enough to ensure that this is not treated in isolation from the metaphysics of physical modality. The present paper argues, however, that modal monism should stay out of the picture, for it cannot add anything to support such treatment. The reason for this is that modal monism is false.

To be sure, modal monism has its own detractors within the field of analytic metaphysics (see, e.g., Fine, 2005, for an argument in favor of modal pluralism). Here I will construct an argument against modal monism based on a technical result in the metatheory of quantum logic that has implications for our understanding of what quantum mechanics deems physically possible. To show that modal monism is false, I provide a counterexample to Williamson's reduction of physical possibility to metaphysical possibility. That is, I argue that there are propositions that quantum mechanics deems physically possible, but which are not metaphysically compossible with what in some circumstances are the laws of nature.

I start by providing some details about the quantum logical interpretation, or as I shall call it (following Stairs, 2015) the quantum logical *reconstruction* of quantum mechanics, enough to make intelligible the presentation of a technical result (due to Pavičić & Megill, 1999) that establishes the *non-categoricity* of quantum logic (where a logic will be called categorical with respect to an isomorphism class of structures if and only if all its models are in that class).

---

[1] The point is recalled towards the end of the book: "[T]he methodology of this book is akin to that of a natural science.... The theories are judged partly on their strength, simplicity, and elegance, partly on the fit between their consequences and what is independently known" (Williamson, 2013, p. 423).

A Note on Williamson's Modal Monism

Afterwards I discuss the connection between this metatheoretical result and what quantum mechanics reveals about modality, suggesting an account of physical possibility that is independent of metaphysical possibility. Then I come back to modal monism to explain why I believe it is false, but also to reject an objection that might be raised against my argument, an objection that insists against taking the laws of quantum logic as laws of nature. An elaboration and defense of modal pluralism in the context of quantum mechanics will, however, be deferred to another paper.

2 Quantum logic and orthomodular lattices

This section briefly introduces the standard approach to quantum logic and its semantics constructed in terms of orthomodular lattices. In doing so, it assumes the standard Hilbert space formalism of quantum mechanics. Thus, as is usual, the pure states of a quantum system are taken to be represented by unit vectors in an associated Hilbert space that represents the state space of the system, and quantum properties to be represented by linear closed subspaces of the Hilbert space.[2]

Let $\mathcal{QL}$ be a formal language that contains an infinite set of formulas, $p, q, \ldots$, and three symbols for logical connectives, $\sim, \wedge, \vee$ (that is, for negation, conjunction, and disjunction, respectively) such that if p and q are sentences in $\mathcal{QL}$, then $\sim p$, $p \wedge q$, and $p \vee q$ are also in $\mathcal{QL}$.[3] The semantics of this formal language is typically given by an ortholattice, which can be defined as an algebraic structure $\mathcal{LA}$, i.e., a set of elements, $a, b, \ldots$ (which are precisely the linear closed subspaces of a Hilbert space) together with the operations $', \cap, \cup$ (for orthocomplementation, join, and meet, respectively), such that any elements a, b, c in that set satisfy conditions like the following:

$$a \cap b = b \cap a$$
$$a \cap (b \cap c) = (a \cap b) \cap c$$
$$a \cap (a \cup b) = a$$
$$a = a''$$
$$(a \cap a') \cap b = a \cap a'$$
$$a \cup b = (a' \cap b')'$$

---

[2] For reference, see any standard introduction to quantum mechanics and its logical-algebraic formalism, e.g., (Beltrametti & Cassinelli, 1981).

[3] One can also define a variety of implication connectives. For a brief review, see (Pavičić, 2016). The first to define different quantum implications was Weyl (1940).

Iulian D. Toader

One further defines a supremum $1 := a \cup a'$, an infimum $0 := a \cap a'$, and a partial order $a \leq b := a \cap b = a$, $a \leq b := a \cup b = b$, as well as an equivalence relation $a \equiv b := (a \cap b) \cup (a' \cap b')$. Then, an ortholattice $\mathcal{LA}$ is said to be a model of $\mathcal{QL}$ if and only if for any sentences p, q in $\mathcal{QL}$ and any elements a, b of $\mathcal{LA}$ there is a map $h : \mathcal{QL} \longrightarrow \mathcal{LA}$, such that the following three conditions are satisfied:

$$h(\sim p) = h(p)' = a'$$
$$h(p \vee q) = h(p) \cup h(q) = a \cup b$$
$$h(p \wedge q) = h(p) \cap h(q) = a \cap b$$

The map h is a homomorphism, i.e., it preserves operational structure, by mapping negation to orthocomplementation, disjunction to join, and conjunction to meet. Thus, the algebraic relations between linear closed subspaces of the Hilbert space (i.e., the equations involving the ortholattice operations) are expressed by sentences of the formal language $\mathcal{QL}$, and are typically taken to represent compatibility relations between the properties of a quantum system (i.e., their co-measurability). But, of course, not all algebraic relations will do so, since in quantum mechanics not all properties of a system are co-measurable. In particular, distributivity, that is, the law according to which, for any elements $a, b, c \in \mathcal{LA}$, $a \cap (b \cup c) = (a \cap b) \cup (a \cap c)$, is usually taken to fail in quantum mechanics. As a result, weaker versions of distributivity have been considered, one of which is modularity.

A lattice is *modular* if and only if the law of modularity holds: if $a \leq c$, then $a \cap (b \cup c) = (a \cap b) \cup (a \cap c)$. As von Neumann first realized, however, modularity requires a finite dimensional Hilbert space, which he thought was improper for a truly general quantum logic.[4] A further weakening of distributivity leads to orthomodularity. A lattice is *orthomodular* if and only if the law of orthomodularity holds: if $a \leq b$, then $b = a \cup (a' \cap b)$. Orthomodular lattices give the standard algebraic semantics of the quantum logical calculus $\mathcal{QL}$. Thus, the homomorphism h, as defined above, maps sentences from $\mathcal{QL}$ to compatibility relations between the elements of an orthomodular structure of quantum properties.

---

[4]This is the reason von Neumann came to give up the Hilbert space formalism several years after he had introduced it. See the quotation at the end of the paper. For a detailed discussion, see, e.g., (Rédei, 1996).

3 Non-categoricity and physical possibility

Having described these basic facts about quantum logic and its algebraic semantics, I turn now to a (so far unduly neglected) result in the metatheory of quantum logic, which shows that quantum logic is non-categorical (Pavičić & Megill, 1999, 2009). More precisely, it is non-categorical with respect to the isomorphism class of orthomodular lattices, because one can show that not all its algebraic models are in this class: "one of its models is an orthomodular lattice, while others are nonorthomodular lattices" (Pavičić, 2016, p. 2).

Nonorthomodular lattices are those in which the law of orthomodularity fails. Some of these nonorthomodular lattices, however, obey a weakened form of orthomodularity, appropriately called *weak orthomodularity*. To see what the law of weak orthomodularity states, consider an alternative definition of orthomodularity: a lattice is orthomodular if and only if for any elements $a, b \in \mathcal{LA}$, $a \equiv b = 1 \Rightarrow a = b$. That is, two elements of an orthomodular lattice are equivalent (in the sense of equivalence defined above) only if they are the same element. A weakly orthomodular, nonorthomodular lattice is then one in which for any $a, b, c \in \mathcal{LA}$, $a \equiv b = 1 \Rightarrow (a \cup c) \equiv (b \cup c) = 1$. In other words, such a lattice includes distinct elements that are indiscernible via orthomodularity.[5]

What Pavičić and Megill have proved, more precisely, is that there exists an orthomodular lattice, as well as a weakly orthomodular, non-orthomodular lattice, such that the quantum logical propositional calculus is sound and complete with respect to both.[6] One interpretation, proposed by Pavičić, is that quantum logic "can simultaneously describe distinct realities" (Pavičić, 2016, p. 2). What this means, I take it, is that quantum logic can associate distinct compatibility structures of properties with one and the same quantum system: both an orthomodular structure, and a weakly orthomodular, nonorthomodular one.

---

[5]It might help to compare this to first-order Peano arithmetic: a nonstandard model includes distinct elements—nonstandard numbers—that are indiscernible by the application of the rules of the theory within its language.

[6]Related claims concerning the semantics of quantum logic were expressed first by Weyl, who pointed out that Birkhoff and von Neumann's quantum logic allows non-unique valuations of disjunctive and conjunctive formulas (Weyl, 1940). For discussion, see (Toader, 2020b). See (Hellman, 1980) for a proof that quantum logical connectives are not truth-functional. For an application of the non-truth-functional semantics developed by Arnon Avron and his collaborators (see, e.g., Avron & Zamansky, 2011) to the quantum logical formalism, see (Jorge & Holik, 2020).

Iulian D. Toader

This raises a couple of questions, which I better attempt to answer now before I even formulate my argument against modal monism. First, one may point out that non-categoricity is an artefact of the quantum logical reconstruction of quantum mechanics, rather than a logical feature of the physical theory itself. Secondly, one may insist that one should reject a non-categorical theory as possessing an undesirable metatheoretical property.

To the first question: one should recall that the purpose of the quantum logical reconstruction is to reveal the logical structure of quantum mechanics—the very objective stated by Birkhoff and von Neumann already in 1936: "to discover what logical structure one may hope to find in physical theories which, like quantum mechanics, do not conform to classical logic" (Birkhoff & von Neumann, 1936, p. 823; see also Suppes, 1966). Thus, to the extent that quantum logic succeeded in at least approximating the logical structure of quantum mechanics, and there is no serious reason to doubt that this is the case, its non-categoricity entails that quantum mechanics associates distinct compatibility structures of properties with a quantum system. In other words, quantum mechanics tells us that physical reality can be either an orthomodular ortholattice or a weakly orthomodular, nonorthomodular one. Despite the fact that quantum logic cannot itself be considered a physical theory, it is nevertheless physically salient because one of its metatheoretical properties has implications for our understanding of what quantum mechanics deems physically possible.[7]

To the second question: that one should characterize a non-categorical theory as semantically defective, as possessing an undesirable metatheoretical property, seems justified. Witness the many attempts to prove that theories like arithmetic and set theory are categorical, on the assumption that non-categoricity is a liability, one that implies for example the semantic indeterminacy of the languages of these theories, their inability to pick out their intended semantics.[8] However, from a metatheoretical perspective, there is arguably a significant difference between non-physical theories and physical ones like quantum mechanics. The categoricity of the latter has been long rejected as a kind of undesirable rigidity that unduly constrains its applicability.[9]

---

[7] The categoricity problem of quantum mechanics is discussed at length in (Toader, 2020a)

[8] See, for discussion, (Button & Walsh, 2018).

[9] Consider, e.g., the following view: "A categorical theory is one such that any two models (true interpretations) of its underlying abstract formalism are isomorphic (structurally identical). Now a necessary condition for theory isomorphism is that the corresponding sets be similar, i.e., that there be a one-to-one correspondence between them. But we do not want such a rigidity in

A Note on Williamson's Modal Monism

More recently, non-categoricity appears to be considered as a theoretical advantage of a physical theory: "A theory that underdetermines its own interpretation is like a healthy breeding population: it has a shot at enough diversity to ... meet the variety of demands its scientific environment places on it" (Ruetsche, 2011, p. 355). One sometimes speaks, very aptly, of "intended non-categoricity" (Rédei, 2014, p. 80) to characterize the intention of the physicist to construct physical theories that allow for non-isomorphic models.

One fundamental idea underlying these evaluations of non-categoricity is that the non-isomorphic models of a physical theory represent distinct physical possibilities. This suggests a semantic account of physical modality, according to which a proposition is physically possible if and only if it is true in at least one model of a non-categorical physical theory. This idea, I submit, should be understood as the basis for an independent account of physical possibility, i.e., one that does not reduce physical possibility to another kind of possibility, like metaphysical possibility.

4 Against Modal Monism

I think that everything is now in place for a presentation of my argument against Williamson's modal monism, an argument that draws, as already announced, on the non-categoricity of the quantum logical formalism. So let p be a sentence in $\mathcal{QL}$ that expresses a proposition $h(p)$ which holds in a weakly orthomodular, nonorthomodular structure of quantum properties, and let's assume that $h(p)$ entails weak orthomodularity, but not the negation of orthomodularity. Similarly, let q be a sentence in $\mathcal{QL}$ that expresses a proposition $h(q)$ which holds in an orthomodular structure of quantum properties, and let's assume that $h(q)$ entails orthomodularity.

According to the independent account of physical modality, sketched above, each of $h(p)$ and $h(q)$ is physically possible, for each is true in at least one structure, i.e., $h(p)$ is true in a weakly orthomodular, nonorthomodular lattice, and $h(q)$ is true in an orthomodular lattice. This situation obtains because, as argued above, quantum logic is non-categorical, i.e., these two mutually non-isomorphic ortholattices are provably in the class of its models. From a quantum-mechanical point of view then, there is no reason to dismiss either $h(p)$ or $h(q)$ as a physical impossibility.

---

physics for, even if two theories do have formally identical basic formulas (e.g., wave equations), they may refer to entirely different kinds of physical systems, these kinds being conceptualised as sets that need not be similar" (Bunge, 1973, p. 166).

Iulian D. Toader

But recall that, according to Williamson's modal monism, that is physically possible which is metaphysically compossible with what in some circumstances are the laws of nature. So take the following circumstances: an orthomodular structure of quantum properties—an orthomodular world, as it were. On Williamson's view, then, $h(q)$ would be physically possible, since it is logically consistent, and thus metaphysically compossible, with the laws of an orthomodular world: by assumption, $h(q)$ entails orthomodularity. Furthermore, $h(p)$ would be physically possible as well, since it is logically consistent, and thus metaphysically compossible, with the laws of an orthomodular world: by assumption, $h(p)$ does not entail the negation of orthomodularity.

So far, so good. Nevertheless, consider now the following, different circumstances: a weakly orthomodular, nonorthomodular structure of quantum properties—a weakly orthomodular, nonorthomodular world, that is. On Williamson's view, $h(p)$ would be physically possible, since it is logically consistent, and thus metaphysically compossible, with the laws of a weakly orthomodular, nonorthomodular world: by assumption, $h(p)$ entails weak orthomodularity. However, on the same view, $h(q)$ would turn out to be physically impossible, because $h(q)$ is not metaphysically compossible with the laws of a weakly orthomodular, nonorthomodular world, since their conjunction is logically inconsistent: by assumption, $h(q)$ entails orthomodularity.

But how can $h(q)$ be physically possible? According to the independent account of physical possibility suggested above, which draws on noncategoricity, metaphysical compossibility with the laws of a weakly orthomodular, nonorthomodular world is neither necessary, nor sufficient for $h(q)$ to be physically possible. What is necessary and sufficient for $h(q)$ to be physically possible is that it be true in at least one of the non-isomorphic models of the theory. Physical modality is dependent on the semantic properties of the physical theory, rather than on metaphysical modality. The fact that $h(q)$ entails orthomodularity renders it false in a nonorthomodular world, but not physically impossible.

It follows from all this that there are propositions which quantum mechanics deems physically possible, but which according to Williamson's modal monism are not metaphysically compossible with what, in some circumstances, are the laws of nature. This, I think, constitutes a sufficient reason for rejecting the reduction of physical possibility to metaphysical possibility, which characterizes modal monism, at least in the version that appears to be advocated by Williamson. Furthermore, insofar as it is actually

A Note on Williamson's Modal Monism

based on modal monism, his anti-exceptionalism about the metaphysics of metaphysical modality turns out to be improperly justified.

One may object that orthomodularity and weak orthomodularity cannot be considered laws of nature (Mittelstaedt, 2012). If so, the above would not really provide a counterexample to Williamson's reduction of physical possibility to metaphysical possibility. This objection raises, of course, a more general question: what are the laws of nature according to quantum mechanics? Typically, laws of nature are supposed to express relations between the states of a physical system, relations that allow us to make testable predictions about the states of a system. So the laws of nature are typically dynamical laws, as they describe trajectories through the state space associated with a system. But in quantum logic, of course, there are no dynamical laws, there is no expression of Schrödinger's equation.

This may be considered as a shortcoming of the quantum logical reconstruction of quantum mechanics, but only if one assumes that any reconstruction must describe the dynamics. This, however, has never been the purpose of quantum logic. As mentioned above, that purpose was to uncover the logical structure of quantum mechanics, the compatibility structure of properties possessed by a quantum system. But a dynamical law does not tell us anything about exactly what quantum properties are compatible (i.e., co-measurable). In the quantum logical reconstruction, as von Neumann emphasized in a famous letter, the focus is not on the states of a system any more, but on its properties: "I would like to make a confession which may seem immoral: I do not believe absolutely in Hilbert space any more.... Now we begin to believe, that ... it is not the *vectors* which matter but the *lattice of all linear (closed) subspaces*. Because: ... the *states* are merely a derived notion, the primitive (phenomenologically given) notion being the *qualities*, which correspond to the *linear closed subspaces*." (von Neumann to Birkhoff, November 6, 1935)

Orthomodularity and weak orthomodularity are laws of nature in the sense that they describe compatibility relations between quantum properties. These are not dynamical laws, since they do not describe trajectories through state space, but they allow us to predict what properties of a system are compatible in certain circumstances. The law of orthomodularity, for example, tells us that no measurement undertaken in an orthomodular world can reveal an instance of an nonorthomodular compatibility structure of properties. It can furthermore be easily seen that such laws support counterfactuals. Thus, the difference between orthomodular and weakly orthomodular, nonorthomodular worlds is not merely factual, but nomological. So the objection fails.

References

Avron, A., & Zamansky, A. (2011). Non-deterministic semantics for logical systems. In D. Gabbay & F. Guenthner (Eds.), *Handbook of Philosophical Logic* (pp. 227–304).
Beltrametti, E. G., & Cassinelli, G. (1981). *The Logic of Quantum Mechanics*. Addison-Wesley.
Birkhoff, G., & von Neumann, J. (1936). The logic of quantum mechanics. *Annals of Mathematics*, 823–843.
Bunge, M. (1973). *Philosophy of Physics*. Dordrecht.
Button, T., & Walsh, S. (2018). *Philosophy and Model Theory*. Oxford University Press.
Fine, K. (2005). The varieties of necessity. In *Modality and Tense* (pp. 235–260). Oxford University Press.
Hellman, G. (1980). Quantum logic and meaning. *Proceedings of the Philosophy of Science Association*, 493–511.
Jorge, J. P., & Holik, F. (2020). Non-deterministic semantics for quantum states. *Entropy*, doi:10.3390/e22020156.
Mittelstaedt, P. (2012). Are the laws of quantum logic laws of nature? *Journal for General Philosophy of Science*, 215–222.
Pavičić, M. (2016). Classical logic and quantum logic with multiple and common lattice models. *Advances in Mathematical Physics*, 1–12.
Pavičić, M., & Megill, N. D. (1999). Non-orthomodular models for both standard quantum logic and standard classical logic: repercussions for quantum computers. *Helvetica Physica Acta*, 189–210.
Pavičić, M., & Megill, N. D. (2009). Is quantum logic a logic? In K. Engesser, D. M. Gabbay, & D. Lehmann (Eds.), *Handbook of Quantum Logic and Quantum Structures: Quantum Logic* (pp. 23–48).
Rédei, M. (1996). Why John von Neumann did not like the Hilbert space formalism of quantum mechanics (and what he liked instead). *Studies in the History and Philosophy of Modern Physics*, 1309–1321.
Rédei, M. (2014). Hilbert's 6th problem and axiomatic quantum field theory. *Perspectives on Science*, 80–97.
Ruetsche, L. (2011). *Interpreting Quantum Theories: The Art of the Possible*. Oxford University Press.
Stairs, A. (2015). Quantum logic and quantum reconstruction. *Foundations of Physics*, 1351–1361.
Suppes, P. (1966). The probabilistic argument for a nonclassical logic of quantum mechanics. *Philosophy of Science*, 14–21.

Toader, I. D. (2020a). The categoricity of quantum mechanics. *unpublished manuscript.*
Toader, I. D. (2020b). Why did Weyl think that quantum logic is a formal pottage? *unpublished manuscript.*
Weyl, H. (1940). The ghost of modality. In M. Farber (Ed.), *Philosophical Essays in Memory of Edmund Husserl* (pp. 278–303). Harvard University Press.
Williamson, T. (2013). *Modal Logic as Metaphysics.* Oxford University Press.
Williamson, T. (2016). Modal science. *Canadian Journal of Philosophy*, 453–492.

Iulian D. Toader
University of Salzburg
Austria
E-mail: `Iulian.Toader@sbg.ac.at`

www.ingramcontent.com/pod-product-compliance
Lightning Source LLC
Chambersburg PA
CBHW062200080426
42734CB00010B/1758